# SCIENCEPLUS®

## TECHNOLOGY AND SOCIETY

LEVEL GREEN

# TEACHING RESOURCES

## Unit 8

Our Changing Earth

HOLT, RINEHART AND WINSTON
*Harcourt Brace & Company*
Austin • New York • Orlando • Atlanta • San Francisco • Boston • Dallas • Toronto • London

# To the Teacher

This booklet contains a comprehensive collection of teaching resources. You will find all of the blackline masters that you need to plan, implement, and assess this unit. Also included are worksheets that correspond directly to the SourceBook.

Choose from the following blackline masters to meet your needs and the needs of your students:

- **Home Connection** consists of a parent letter designed to get parents involved in the excitement of the *SciencePlus* method. The letter provides parents with a general idea of what you are going to cover in the unit, and it even gives you an opportunity to ask for any common household materials that you may need to accomplish the unit's activities most economically.
- **Discrepant Event Worksheets** provide demonstrations and activities that seem to challenge logic and reason. These worksheets motivate students to question their previous knowledge and to develop reasonable explanations for the discrepant phenomena.
- **Math Practice Worksheets** and **Graphing Practice Worksheets** help fine-tune math and graphing skills.
- **Theme Worksheets** encourage students to make connections among the major science disciplines.
- **Spanish Resources** include Spanish versions of the Home Connection letter, plus worksheets that outline the big ideas and principal vocabulary terms for the unit.
- **Transparency Worksheets** correspond to teaching transparencies to help you reteach, extend, or review major concepts.
- **SourceBook Activity Worksheets** reinforce content introduced in the SourceBook.
- **Resource Worksheets** consist of blackline-master versions of charts, graphs, and activities in the Pupil's Edition.
- **Exploration Worksheets** consist of blackline-master versions of Explorations in the Pupil's Edition. To help students focus on specific tasks, many of these worksheets include a goal, step-by-step instructions, and even cooperative-learning strategies. These worksheets simplify the tasks of assigning homework, allowing opportunities for make-up work, and providing lesson plans for substitute teachers.
- **Unit Activity Worksheet** consists of an activity, such as a crossword puzzle or word search, that provides a fun way for students to review vocabulary and main concepts.
- **Review Worksheets** consist of blackline-master versions of the review materials in the Pupil's Edition, including Challenge Your Thinking, Making Connections, and SourceBook Unit CheckUp.
- **Chapter Assessments** and **End-of-Unit Assessments** provide additional assessment questions. Each assessment worksheet includes two or more Challenge questions that encourage students to synthesize the main concepts of the chapter or unit and to apply them in their own lives.
- **Activity Assessments** are activity-based assessment worksheets that allow you to evaluate students' ability to solve problems using the tools, equipment, and techniques of science.
- **Self-Evaluation of Achievement** gives you an easy method of monitoring student progress by allowing students to evaluate themselves.
- **SourceBook Assessment** is an easy-to-grade test consisting of multiple-choice, true-false, and short-answer questions.

For your convenience, an **Answer Key** is available in the back of this booklet. The key includes reduced versions of all applicable worksheets, with answers included on each page.



Credits: See page 92.

SCIENCEPLUS is a registered trademark of Harcourt Brace & Company licensed to Holt, Rinehart and Winston, Inc.

Printed in the United States of America

ISBN 0-03-095666-8 2 3 4 5 6 7 8 9 021 99 98 97 96

# Unit 8: Our Changing Earth

## Contents



▼ *A corresponding transparency is available. See the Teaching Transparencies Cross-Reference on the next page.*

## Contents, continued



## Teaching Transparencies Cross-Reference

| Transparency | Corresponding Worksheet | Suggested Point of Use |
|---|---|---|
| **Chapter 24** | | |
| 78: Clues to Continental Drift | Transparency Worksheet, p. 7 | Lesson 2, Putting the Pieces Together, p. 460 |
| 79: Plate Movement and Volcanoes | Transparency Worksheet, p. 11 | Lesson 2, Continental Facts, p. 464 |
| 80: The Seven Major Plates | | Lesson 2, The Seven Major Plates and Their Movement, p. 465 |
| 81: Breakup of Pangaea | | Lesson 2, Five Stages in the Breakup of Pangaea, p. 467 |
| **Chapter 25** | | |
| 82: Collecting Clues in the Schoolyard | Exploration Worksheet, p. 23 | Lesson 1, Exploration 1, p. 473 |
| 83: Gauging the Change | | Lesson 1, Gauging the Change, p. 476 |
| 84: The Niagara Story | | Lesson 1, The Niagara Story, p. 477 |
| 85: Investigating Rock-Breaking Forces | Exploration Worksheet, p. 26 | Lesson 2, Exploration 3, p. 482 |
| **Chapter 26** | | |
| 86: The Water Cycle | Transparency Worksheet, p. 38 | Lesson 1, The Power of Water, p. 494 |
| 87: Water Below! | | Lesson 1, Water Below! p. 497 |
| 88: Glaciers | Transparency Worksheet, p. 48 | Lesson 6, Glaciers—Rivers of Ice, p. 512 |
| 89: Glacier on the Move | | Lesson 6, Glaciers on the Move, p. 514 |
| 90: Geology Match-Up | Unit Activity Worksheet, p. 61 | End of Unit 8 |

Dear Parent,

In the next few weeks, your son or daughter will learn about the Earth's history. The students will explore how the landscape of the Earth has changed and some of the forces that have caused these changes. They will also investigate the effects that water, wind, and ice have on the land. By the time the students have finished Unit 8, they should be able to answer the following questions to grasp the "big ideas" of the unit.

1. What is geology all about? (Ch. 24)
2. What does the expression "the present is the key to the past" mean? (Ch. 24)
3. How do geologists use inferences and observations? (Ch. 24)
4. What is some evidence that supports the theory of continental drift? (Ch. 24)
5. What are the main features of the Earth's interior? (Ch. 24)
6. Use your knowledge of the structure of the Earth to explain why the floor of the Atlantic Ocean is growing wider. (Ch. 24)
7. What are some of the agents of weathering? (Ch. 25)
8. What role does water play in the weathering process? (Ch. 25)
9. What role does gravity play? (Ch. 25)
10. What is some evidence for widespread glaciation in the past? (Ch. 26)

You may want to ask your son or daughter some of these questions as we progress through the unit. This will give him or her valuable practice communicating his or her knowledge. Don't worry if the answers do not sound like they were memorized. With this program, students discover the answers for themselves either individually or in groups. That's both the challenge and the fun of science.

Sincerely,

The items listed below are materials that we will use in class for the various science explorations of Unit 8. Your contribution of materials would be very much appreciated. I have checked certain items below. If you have these items and are willing to donate them, please send them to the school with your child by

______________________________________.

- ◯ buckets
- ◯ candles
- ◯ cardboard (thin sheets)
- ◯ cardboard tubes
- ◯ clothespins or small clamps
- ◯ containers (identical and transparent)
- ◯ food coloring
- ◯ jar lids
- ◯ jars (large)
- ◯ latex gloves
- ◯ magazines
- ◯ magnifying glasses
- ◯ markers
- ◯ modeling clay
- ◯ newspapers
- ◯ pans (shallow, rectangular)
- ◯ paper towels
- ◯ pitchers
- ◯ plywood, about 20 cm × 30 cm
- ◯ poster board
- ◯ rock samples (granite, limestone, sandstone, brick, pieces of concrete)
- ◯ rubber tubing
- ◯ sand
- ◯ soil samples (sand, clay, or garden)
- ◯ trays (large, plastic or metal)
- ◯ wax pencils
- ◯ wire
- ◯ wire cutters
- ◯ wooden blocks (large)

Thank you in advance for your help.

# Earth Trivia

Complete this activity as you read Lesson 1, Seeing the Sights, which begins on page 453 of your textbook.

Check your knowledge of Earth facts by answering the questions below. You may find certain resources such as encyclopedias or almanacs helpful in your search.

**1.** The longest river in the world is 6670 km long. It is the

**a.** Mackenzie in Canada. **b.** Nile in Africa. **c.** Mississippi in the United States.
**d.** Chang in Asia. **e.** Amazon in South America.

**2.** Mount Everest is the highest mountain peak in the world. Its height above sea level is approximately

**a.** 3 km. **b.** 5 km. **c.** 7 km. **d.** 9 km. **e.** 11 km.

**3.** The deepest place on Earth is the Marianas Trench in the Pacific Ocean. Its depth below sea level is

**a.** 3 km. **b.** 5 km. **c.** 7 km. **d.** 9 km. **e.** 11 km.

**4.** As of 1992, the tallest free-standing structure in the world was the CN Tower in Toronto, Canada. It is 550 m (more than 0.5 km) high. The highest natural waterfall in the world is Angel Falls in Venezuela. It is __________ the height of the CN Tower.

**a.** one-third **b.** one-half **c.** equal to **d.** two times **e.** three times

**5.** The length of the Rocky Mountains, which is the second longest range in the world, is

**a.** 2000 km. **b.** 4000 km. **c.** 6000 km. **d.** 8000 km. **e.** 10,000 km.

**6.** The search for coal, natural gas, and oil leads us to dig deep into the Earth. How deep is the deepest mine, Western Deep in South Africa?

**a.** 0.5 km **b.** 1 km **c.** 2 km **d.** 3 km **e.** 4 km

**7.** North America is a large continent. But how large is it? Among all continents in the world, North America ranks __________ in area.

**a.** first **b.** second **c.** third **d.** fourth **e.** fifth

**8.** Our planet is called "Earth," not "Water." Should it be? The surface is covered by both land and water. Which is the correct proportion of the two?

**a.** The areas of land and water are about equal.

**b.** The area of land is one-half the area of water.

**c.** The area of land is twice the area of water.

**d.** The area of land is three times the area of water.

**9.** The largest freshwater lake in the world (by volume) is in

**a.** the United States. **b.** Canada. **c.** Russia. **d.** Tanzania. **e.** Malawi.

**10.** The largest island in the world is

**a.** Greenland. **b.** New Guinea. **c.** Great Britain.
**d.** Victoria Island (Canada). **e.** Ellesmere Island (Canada).

Name ______________________ Date ____________ Class ____________

# EXPLORATION 1

**Chapter 24**
Exploration Worksheet

## Selling Scenery, page 454

### Cooperative Learning Activity

| | |
|---|---|
| **Group size** | 3 to 4 students |
| **Group goal** | to write a sales presentation or design a travel brochure for geologists |
| **Individual responsibility** | Each member of your group should choose a role such as artistic director, copywriter, team manager, or editor. |
| **Individual accountability** | Each group member should write a self-assessment to review his or her personal participation in the group and the effectiveness of his or her contributions. |

How good a salesperson are you? Test your skill. Choose *one* of the activities below.

1. You are a travel agent. What would you say to your clients to persuade them to visit the scenic landscape you chose in step 4 on page 453 of your textbook? Write down the main points you would make.
2. Design a travel brochure to persuade vacationers to visit the area you chose. Highlight the area's scenic attractions.
3. You are a travel agent again, but this time your client is a geologist. How would you change your sales pitch?
4. Now you are a geologist, and you are planning an expedition to a place of your choice. Where would you go? What would be your sales pitch to get others to join you?

Pitch your strategy in the space below. Feel free to use a separate piece of paper or any classroom materials to enhance your plans.

## A Continental Jigsaw Puzzle, page 459

| Your goal | to determine how all of the continents may once have fit together to form Pangaea |
|---|---|

Imagine that you are a young geologist living in Germany in the early 1900s. Looking at a map of the coasts of Africa and South America, you notice that the coastlines appear to fit together like the pieces of a jigsaw puzzle. You also remember reading about the discovery of fossils of a small lizard called *Mesosaurus* that were found only in eastern South America and southwestern Africa. You then begin to wonder whether all the continents can be made to fit together.

Cut out the continents from the map on the next page. Then try to put the pieces together to get the best possible fit. What clues do you have? Are there missing pieces?

Finally, you believe you have found the right fit. You ask yourself, "What does this mean? How can I possibly explain what I think happened in the past?" For days you puzzle over the answer. At last you think you have worked out a possible solution. With great excitement you write a letter to the Royal Geological Society in London and ask them to consider your ideas. What might the letter say? Finish the letter started here, and sign your name as Alfred Wegener.

*April 1, 1911*

*Royal Geological Society*
*Holywell House, Worship Street*
*London, EC2A2EN*

*My Esteemed Colleagues,*

*I wish to bring to your attention a matter of great importance. For several months I have been working on an idea . . .*

Letter also on page 459 of your textbook

Name ______________________ Date ____________ Class ____________



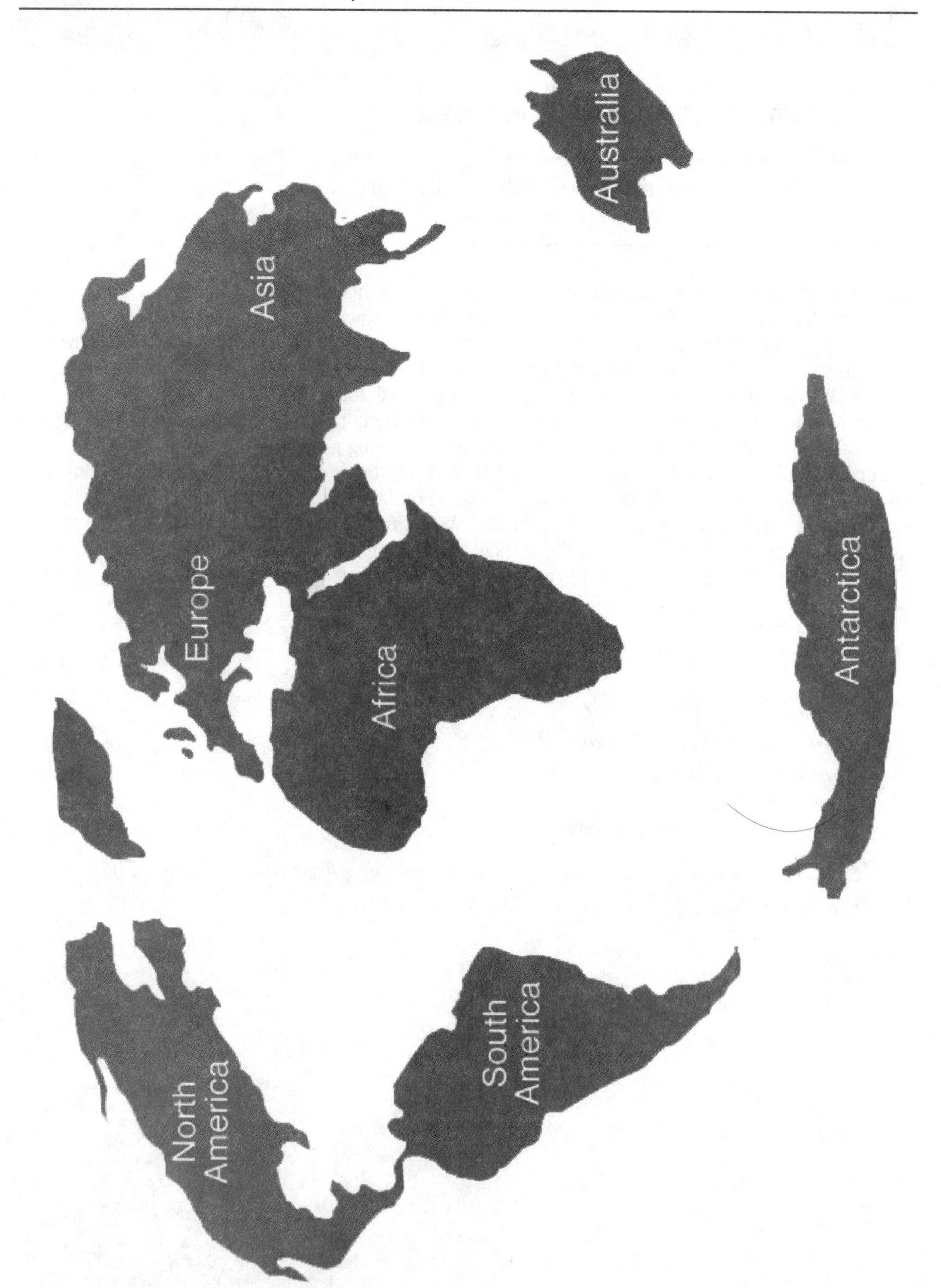

Name ________________________ Date ____________ Class ____________

**Chapter 24**
Transparency Worksheet

Chapter 24

## Clues to Continental Drift — Teacher's Notes

This worksheet corresponds to Transparency 78 in the Teaching Transparencies binder.

---

**Suggested Uses**

Use as a visual aid with either of the topics listed below.

Putting the Pieces Together, page 460
A Further Look at Continental Drift, page 466

Use to discuss and extend the study of the following:

**a.** Wegener's theory of Pangaea

**b.** Wegener's idea of a supercontinent, and the fact that the same rocks and fossils can be found on more than one continent

Use the transparency with the transparency worksheet on the next page for reteaching or review.

---

**Possible Review Question**

What is Wegener's theory of continental drift?

---

**Answer to Review Question**

Wegener's theory of continental drift is based on the belief that all continents were joined together at one time. Over a period of time, according to the theory, some force caused the continents to split apart and drift to their present locations.

Name ______________________ Date ____________ Class ____________

**Chapter 24**
Transparency Worksheet

# Clues to Continental Drift

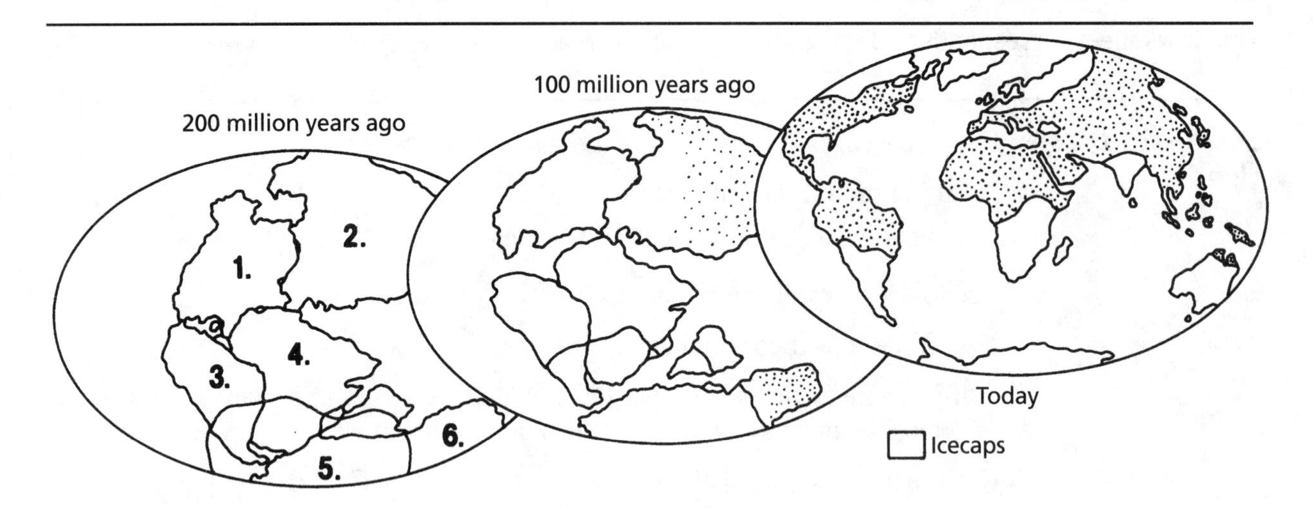


## Glacial Evidence of Continental Drift

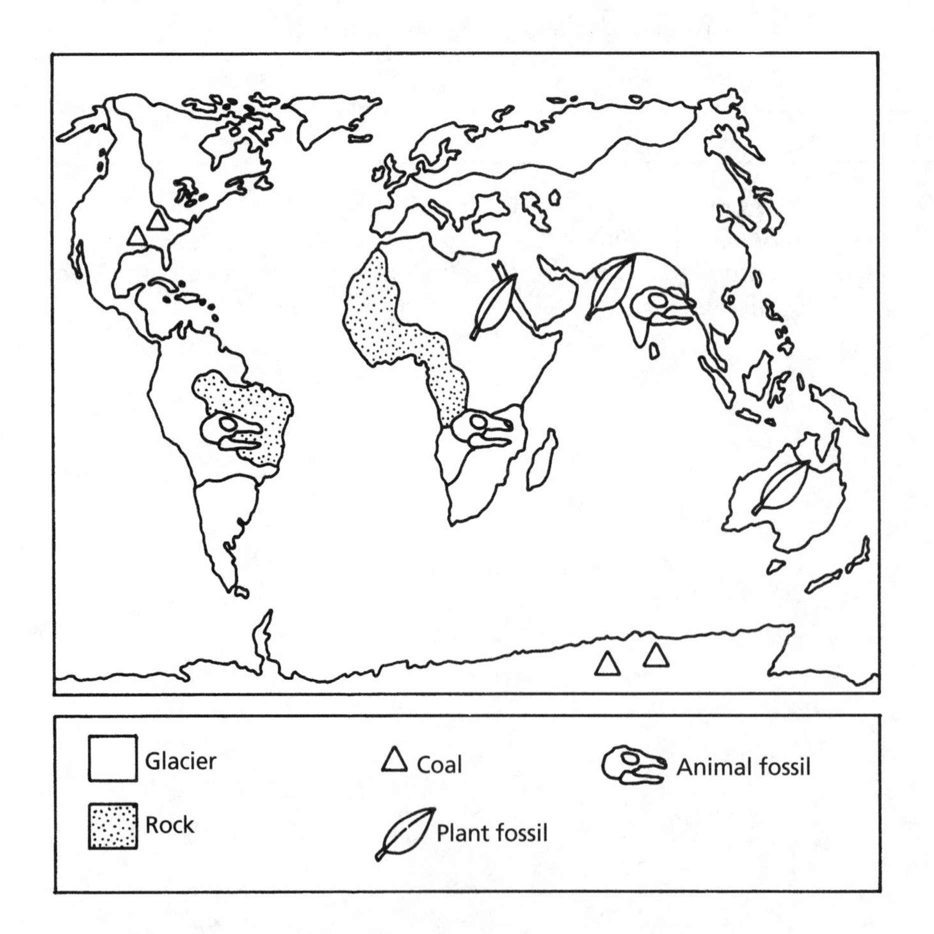


## Fossil Evidence of Continental Drift

Name ______________________ Date __________ Class __________

# EXPLORATION 3

Chapter 24
Exploration Worksheet

Chapter 24

## Creating a Model, page 462

| Your goal | to create a model to study plate movement by convection current | Safety Alert!  |
|---|---|---|

Although questions still remain, most scientists think that plate movement is caused by convection currents. When a gas or a liquid is heated unevenly, the part that is heated rises. The movement of heated gas or liquid is called a *convection current.* To study processes that they cannot see, scientists often make models. In this Exploration, you will create a model of plate movement by convection current.

### You Will Need

- water
- a shallow, rectangular pan (30 cm × 40 cm)
- a hot plate
- dark food coloring
- 8 pieces of thin cardboard, about 1 cm × 1 cm
- a watch or clock with a second hand

### What to Do

1. Fill the pan three-quarters full of water. Place the pan on the hot plate.
2. Heat the water over very low heat for 30 seconds. Add a few drops of food coloring to the center of the pan. Write down your observations in the space provided. Turn off the heat.

_______________________________________________

_______________________________________________

_______________________________________________

_______________________________________________

3. Label the cardboard pieces 1 through 8. Carefully place pieces 1 through 4 in the center of the pan, as close together as possible. Place the rest of the cardboard pieces in the corners of the pan.
4. Turn the hot plate on low. In your ScienceLog, sketch the pattern of movement for each cardboard piece. Turn off the heat.

### Thinking Things Over

1. What happened when you added food coloring to the water? How can you explain your observations?

_______________________________________________

_______________________________________________

_______________________________________________

_______________________________________________

Name ______________________ Date ____________ Class ____________



2. Describe what happened to the cardboard pieces when the water was heated.

______________________________

______________________________

______________________________

______________________________

______________________________

______________________________

Do you think this activity is a good model for plate tectonics? Explain.

______________________________

______________________________

______________________________

______________________________

3. Examine the illustration of the Earth's interior below. What portion of the Earth does each of the following represent: the water, the cardboard pieces, the hot plate?

______________________________

______________________________

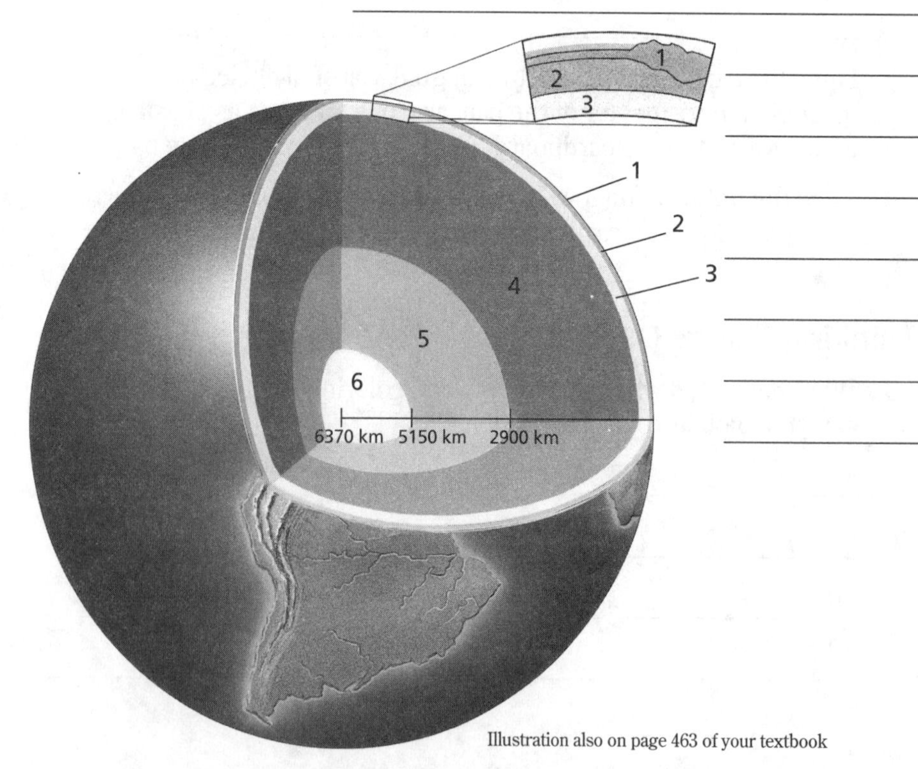


______________________________

______________________________

______________________________

______________________________

______________________________

______________________________

______________________________

Illustration also on page 463 of your textbook

Name ______________________ Date ____________ Class ____________

Chapter 24
Transparency Worksheet



## Plate Movement and Volcanoes — Teacher's Notes

This worksheet corresponds to Transparency 79 in the Teaching Transparencies binder.

### Suggested Uses

Use as a visual aid with the topic listed below.

Continental Facts, page 464

Use to discuss and extend the study of the following:

a. Volcanic eruptions: Explain to students what a volcano is and what happens in a volcanic eruption. Students may also benefit from a brief discussion of the four classifications of volcanic activity: active, intermittent, dormant, and extinct.

b. Plate tectonics: Discuss how tectonic forces cause volcanic activity. Students will also benefit from a discussion of how these same tectonic forces can cause earthquakes.

c. The difference between magma and lava: Magma is hot, melted rock deep within the Earth; lava is magma that emerges on the Earth's surface.

d. Trenches: Discuss how the collision of a continental plate with an oceanic plate can cause a trench to form.

Use the transparency with the transparency worksheet on the next page for reteaching or review. Please note: The worksheet is a reproduction of the actual transparency with certain labels omitted. For answers to that worksheet, see the transparency.

### Possible Extension Questions

1. What is a volcano?
2. How does the collision of a continental plate with an oceanic plate trigger a volcanic eruption?
3. Why do some regions of the world that experience a great deal of volcanic activity also experience many earthquakes?

### Answers to Extension Questions

1. A volcano is an opening in the Earth through which magma comes to the surface. The term also refers to the mountain or cone-shaped structure that is formed by the magma.
2. When the two plates collide, the collision usually drives the oceanic plate under the continental plate. Upon reaching the hotter temperatures of the Earth's interior, the underlying plate starts to melt. This melted rock, or magma, is lighter than the rock around it and therefore starts to rise. This is the beginning of a volcanic eruption.
3. Volcanic activity is often caused by tectonic movement of plates, which can also trigger earthquakes.

Name ______________________ Date ____________ Class ____________

**Chapter 24**
Transparency Worksheet

## Plate Movement and Volcanoes

Name ______________________ Date ____________ Class ____________





## Using the Theme of Energy

This worksheet is an extension of the theme strategy outlined on page 464 of the Annotated Teacher's Edition. It is also an extension of Continental Facts on page 464 of the Pupil's Edition.

| **Focus question** | How is the occurrence of an earthquake along a sliding plate boundary similar to the breaking of a stretched rubber band? | **Safety Alert!**  |
|---|---|---|

### You Will Need

- two wooden blocks, 5 cm × 10 cm × 20 cm
- a rubber band, size 33

### What to Do

1. Lay the two wooden blocks side by side on a desk or the floor.
2. Position the rubber band around both wooden blocks as shown below.

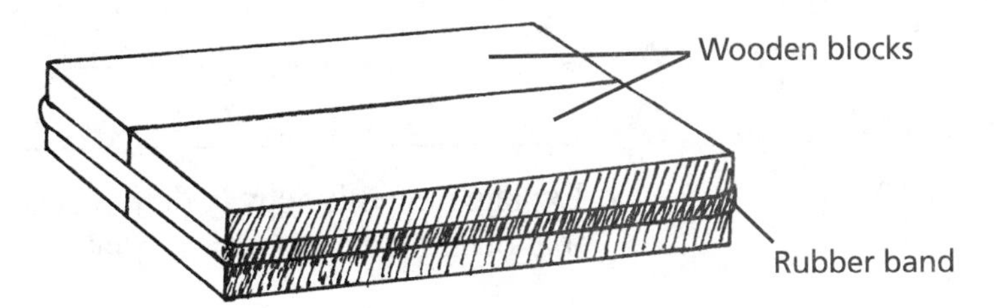


3. Move one of the blocks slowly in one direction while moving the other block in the opposite direction. Describe what happens when the blocks are moved.

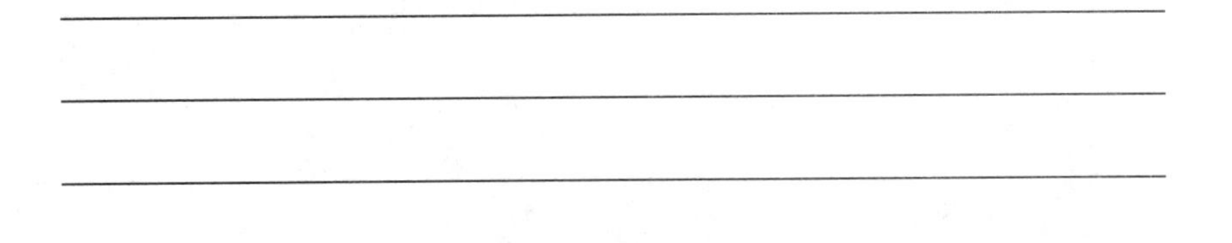

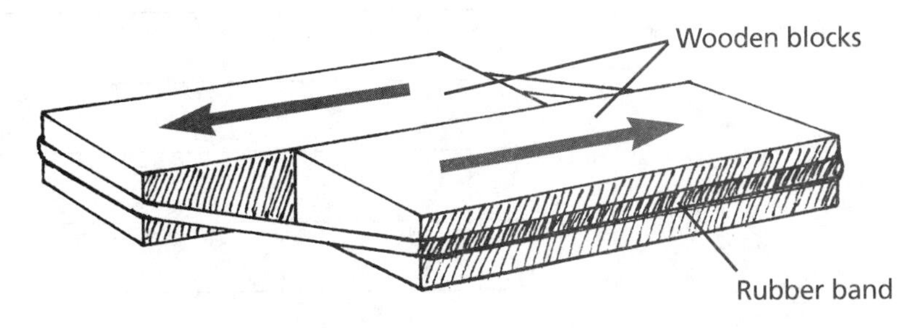

4. Continue moving the blocks in opposite directions, slowly and steadily. Describe what happens.

5. As the rubber band stretches, energy is stored in the rubber band. When is this stored energy released?

6. Explain how this activity demonstrates the buildup and release of energy that occurs during an earthquake at a sliding plate boundary.

Name ______________________ Date ____________ Class ____________

**Chapter 24**
Activity Worksheet

Chapter 24

## Geo-Cross

Complete this activity before beginning Challenge Your Thinking on page 468 of your textbook.

---

Use the clues below to complete the crossword puzzle.

**Across**

1. A small aquatic lizard whose fossils are found in only two places in the world
2. Meaning "all Earth," this word is the name of a supercontinent that may have existed 300 million years ago
7. Plate interactions can cause these.
9. The outermost layer of the Earth; it is thin and rocky.
10. The Earth's center is known as this.
11. 1 down and 9 across together are known as this.
12. The layer of the mantle directly below the lithosphere
15. A scientist who studies the Earth
16. The only major plate that does not have a continent on it

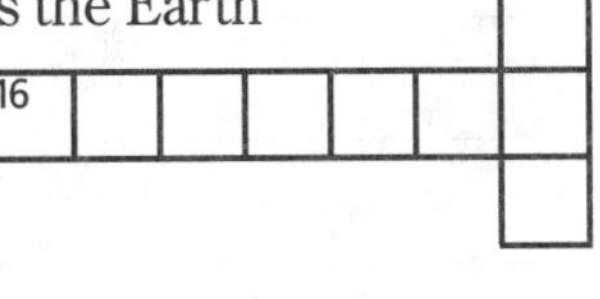

**Down**

1. The middle layer of the Earth, between the crust and the core
2. The Earth's surface is made up of a number of huge _______.
3. These are created when sections of the Earth's surface are folded and thrust upward.
4. These are caused by sections of the Earth's surface sliding past each other, often violently.
5. This man hypothesized the existence of 2 across.
6. The idea that the Earth's surface is made of moving pieces is called the theory of plate _______.
8. This is often expelled from the eruptions of 7 across.
13. A deep valley caused when one section of the Earth's surface is forced downward
14. There are _______ major plates on the Earth's surface.

Name ______________________ Date __________ Class __________

**Chapter 24**
**Review Worksheet**

## Challenge Your Thinking, page 468

### 1. Island Hopping

A *hot spot* is an area of concentrated heat, deep within the Earth's interior. Hot spots melt the rocks of the mantle, which creates a column of magma that rises toward the surface. Examine the map of the Hawaiian Islands below.

**a.** Use the map and your knowledge of the theory of plate tectonics to explain how the Hawaiian Islands might have formed.

______________________

______________________

______________________

______________________

**b.** Which islands are the oldest? How can you tell?

______________________

______________________

______________________

______________________

**c.** Do you think any new islands will form? Why?

______________________

______________________

______________________

______________________

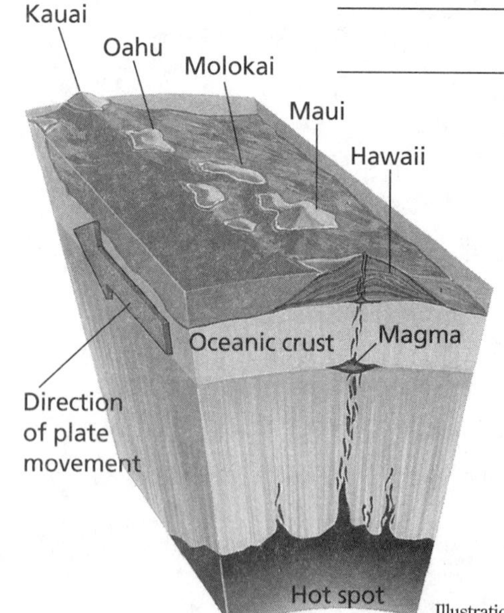


Illustration also on page 468 of your textbook

**d.** Which islands are most likely to have active volcanoes? Explain.

______________________

______________________

______________________

______________________

______________________

Name ______________________ Date __________ Class __________



**2. A More Persuasive Argument** Using the information contained in this chapter, revise the letter that you wrote to the Royal Geological Society in Exploration 2 on page 459 of your textbook. Carefully organize any additional evidence to make your argument as persuasive as possible. Continue in your ScienceLog if necessary.

*April 1, 1911*

*Royal Geological Society*
*Holywell House, Worship Street*
*London, EC2A2EN*

*My Esteemed Colleagues,*

*I wish to bring to your attention a matter of great importance. For several months I have been working on an idea . . .*

Letter also on page 459 of your textbook

Name ______________________ Date ____________ Class ____________



**3. Lizard Debate**

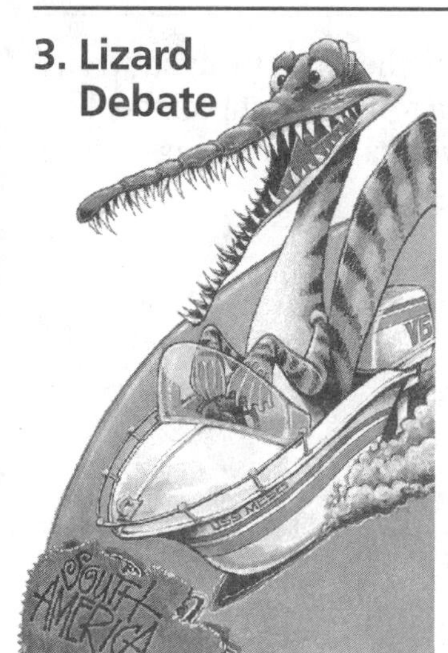


Illustration also on page 468 of your textbook

Fossils of the reptile *Mesosaurus* are found only in South America and Africa. Wegener used this evidence to support his theory of continental drift. Imagine that you are a scientist who refuses to accept Wegener's theory. Suggest other explanations for the location of *Mesosaurus* fossils. They certainly didn't have speedboats! Can you come up with a more reasonable explanation?

______________________________________________

______________________________________________

______________________________________________

______________________________________________

______________________________________________

**4. Photo Mystery!**

Carefully examine the photo at the top of page 469 of your textbook. Make inferences about what caused the scene. How would you check the accuracy of your inferences?

______________________________________________

______________________________________________

______________________________________________

______________________________________________

______________________________________________

**5. Sunken Treasure**

Imagine that you are a travel agent. You are offering submarine tours of a spreading center at the bottom of the ocean. In your ScienceLog, write a brochure or sales pitch designed to convince an average tourist (not a geologist) to take this tour.

Illustration also on page 469 of your textbook

Name ______________________ Date __________ Class __________

Chapter 24
Assessment

Chapter 24

## Correction/ Completion

**1.** The following sentences are incorrect or incomplete. Make the sentences correct and complete.

**a.** The mantle and the crust make up the portion of the Earth known as the asthenosphere.

______________________________________________

______________________________________________

______________________________________________

______________________________________________

**b.** The lithosphere is divided into segments known as ______________ plates.

## Short Responses

**2.** Place a check mark beside each observation that supports the theory of continental drift.

**a.** ______ The continents look somewhat like puzzle pieces that fit together.

**b.** ______ Portuguese is spoken by most people in Brazil and in Portugal.

**c.** ______ Rock formations found in Africa are identical to formations found in South America.

**d.** ______ Scientists have discovered dinosaur fossils that are 200 million years old.

**3.** Match each type of plate activity with a geological event.

**a.** separating ______ Earthquakes occur because of friction between plates.

**b.** converging ______ New crust is formed as molten rock flows into the gap between two plates.

**c.** sliding ______ Trenches are formed in the ocean floor.

Name ____________________ Date ____________ Class ____________



**Short Essay**

4. Explain how deep trenches in the ocean floor and nearby mountain peaks are related to each other.

CHALLENGE

**Word Usage**

5. One definition of *tectonic* is "having to do with building or construction." How does this definition relate to what you've learned about the theory of plate tectonics?

Name ______________________ Date __________ Class __________

## Illustration for Interpretation

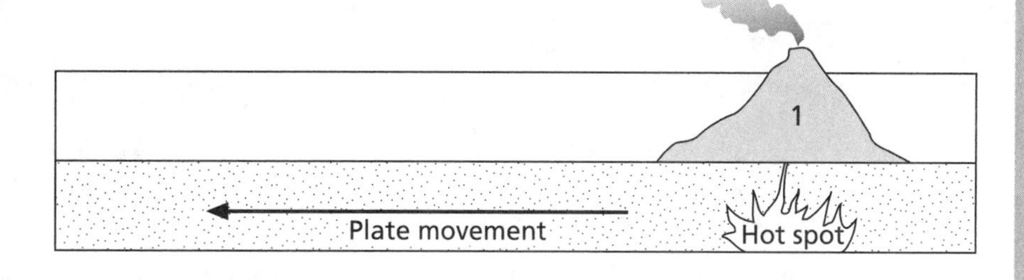


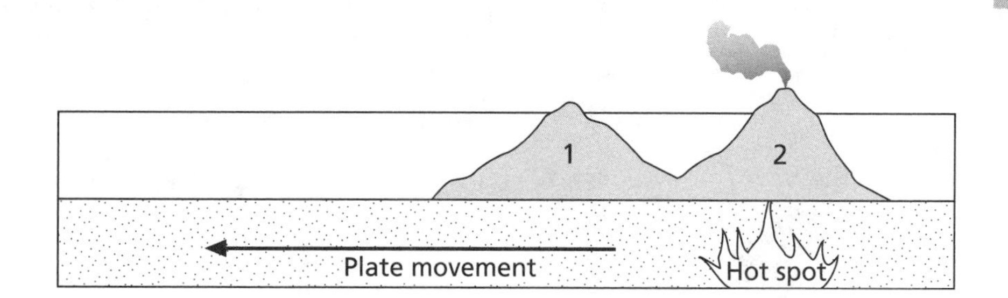


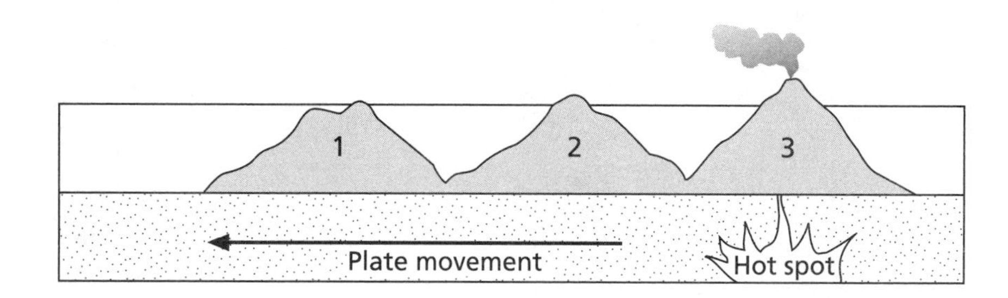


**6. a.** Examine the series of diagrams above and explain what process is described by them.

______________________________________________

______________________________________________

______________________________________________

______________________________________________

______________________________________________

**b.** Which island is the oldest? ______

**c.** Do the diagrams support or disprove the theory of plate tectonics? Explain.

______________________________________________

______________________________________________

______________________________________________

______________________________________________

Name ________________________ Date ____________ Class ____________

Chapter 25
Discrepant Event Worksheet

# Penny Protection

Conduct the following experiment as you begin Chapter 25, which starts on page 470 of your textbook.

## You Will Need

- a large aluminum pie plate
- a hose with a sprayer attachment
- 5 or 6 pennies
- a small mound of pottery clay (do not use modeling clay or other oily clays)

## What to Do

1. Form a flat brick out of the clay, about 15 cm × 5 cm × 3 cm. Place the clay in the aluminum pie plate. Then firmly press the pennies into the top of the clay. Make sure that the pennies are spaced evenly along the top of the clay brick.
2. Take the pie plate outside.
3. What do you predict would happen if you directed a stream of water from the hose at the pennies?

____________________________________________

____________________________________________

4. Now spray the brick with water from the hose. Begin by using a gentle spray and then increase the pressure until the brick begins to wash away.

## Questions

1. What happens to the clay when the hose is turned on?

____________________________________________

____________________________________________

2. What do you observe about the clay underneath the pennies?

____________________________________________

____________________________________________

3. Why do you think this happened?

____________________________________________

____________________________________________

4. Imagine that the pennies are actually a very hard rock, such as granite, and that the clay is a very soft rock such as limestone. In your ScienceLog, draw a diagram showing what the rock formation might look like before and after the effects of the water.

Name ______________________ Date ____________ Class ____________

# EXPLORATION 1

**Chapter 25**
Exploration Worksheet

## Collecting Clues in the Schoolyard, page 473

### Cooperative Learning Activity

| | |
|---|---|
| **Group size** | 3 to 4 students |
| **Group goal** | to find clues about the geological history of your schoolyard, document those clues in the form of a map, and make inferences about the significance of those clues |
| **Individual responsibility** | Each member of your group should choose a role such as secretary, map maker, historian, or primary explorer. |
| **Individual accountability** | Each group member should be able to write a summary of the case-study report and to add his or her comments about the activity. |

Chapter 25

1. Discuss in class how you would help the geologist you read about in Changes in the Schoolyard on pages 471 and 472 of your textbook. Then draw up a preliminary list of clues in your ScienceLog.
2. Go outside to your schoolyard and observe! Add any other clues you think of to your list.
3. Draw a map of your schoolyard in your ScienceLog. Indicate on the map where you found each clue.
4. Is your evidence complete? Do you have enough clues?

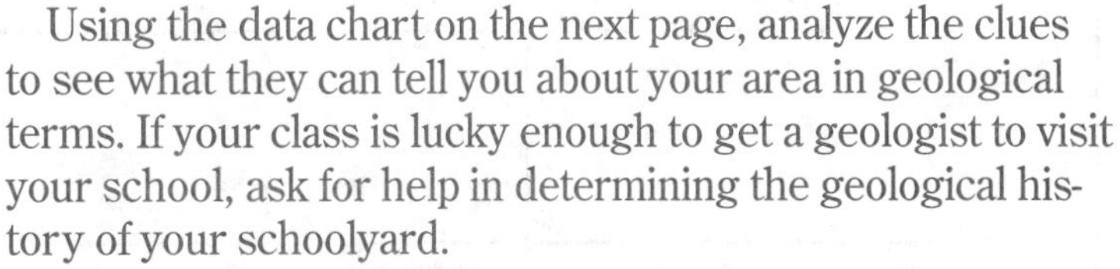

Using the data chart on the next page, analyze the clues to see what they can tell you about your area in geological terms. If your class is lucky enough to get a geologist to visit your school, ask for help in determining the geological history of your schoolyard.

Other classes might be interested in your results, especially after watching your class working outside. In your ScienceLog, write a case-study report to describe what you found out about the geology of your schoolyard and the surrounding area. Share your discoveries with other classes.

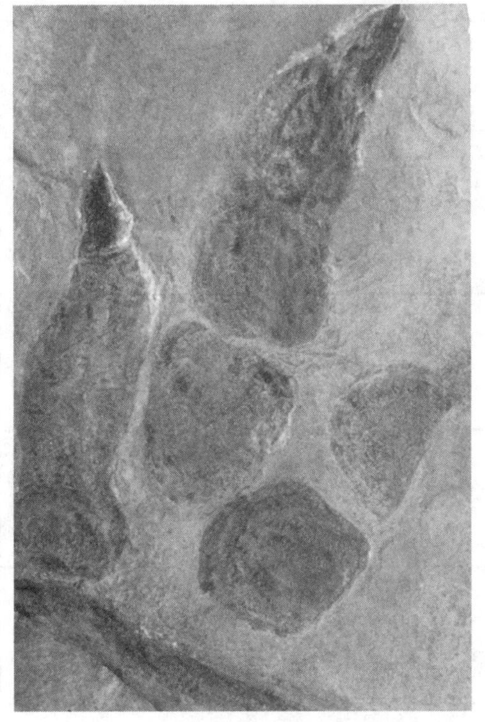

Dinosaur footprint, Connecticut Valley, Massachusetts

Photo also on page 473 of your textbook

Exploration 1 Worksheet, continued

| Clues | Possible history |
|---|---|
| 1. | |
| 2. | |
| 3. | |
| 4. | |
| 5. | |
| 6. | |
| 7. | |
| 8. | |
| 9. | |
| 10. | |
| 11. | |
| 12. | |

Name ______________________ Date ____________ Class ____________

# EXPLORATION 2

Chapter 25
Exploration Worksheet

## A Science Project: Recognizing Change, page 479

| Your goal | to recognize and record the progress of a change taking place in your environment |
|---|---|

Look for a feature in your backyard, on the way to school, or around your town that seems to show a change taking place. It might be a crack in a rock, a spot of bare ground, or a gully formed in a hill by running water.

Once you have found a suitable feature, prove that a change is occurring. The ideas below will help you conduct your project scientifically.

Before you begin your investigation, plan ahead. Observe the condition of the feature over a period of time. Compare its condition when you started your investigation with its condition over time. In your ScienceLog, write down what you see; it's easy to forget details after a while. (Sketches and diagrams often help also.) When you have completed your observations, prepare a report with the following sections:

1. Results of the Investigation (Write down the change you observed.)
2. Investigating the Change (Give details of how you collected your information.)
3. Results Obtained (Include drawings, photos, a data sheet, etc.)
4. Inferences and Summary (What caused the change? Will the change continue to take place? What will eventually result? Can the change be reversed? Can humans have any effect on the change?)

### One Student's Investigation

Caroline decided to watch for a change under a downspout at the corner of her house. First she took a photo and made a drawing of the area directly under the downspout. Then she observed the area closely. She noticed that a change occurred after a rainfall. She made measurements of the changed area. Caroline also made notes about some features of the area: the number of rocks; their appearance, size, and position; and any living things nearby. After each rainfall, she made further observations and measurements. Finally, Caroline summarized her findings in a brief report.

Name ______________________ Date __________ Class __________

# EXPLORATION 3

Chapter 25
Exploration Worksheet

## Investigating Rock-Breaking Forces, page 482

| Cooperative Learning Activity | | Safety Alert! |
|---|---|---|
| **Group size** | 3 to 4 students | Wear safety goggles when chipping or breaking rock.  |
| **Group goal** | to find evidence that various agents of change are at work in familiar areas | |
| **Individual responsibility** | Each member of your group should choose a role such as chief investigator, investigative assistant, reporter, or safety inspector. | |
| **Individual accountability** | Each group member should be able to write a paragraph analyzing how the weathering process in an area he or she examined could be slowed or prevented. | |

Do you know how much force you would need to break a rock? Have you ever tried to do so? Some rocks break easily, but tremendous force is usually required. Though you may not realize it, rock-breaking forces are present in your schoolyard and are acting at this very instant. Think of all the places where there is rock, brick, or concrete. Look at some of them on your way home from school. Can you find evidence that forces of change are at work?

Choose three different locations to investigate. Consider the suggestions below as you carry out your study. Record your observations in your ScienceLog and then summarize your findings in the table on the next page.

### What to Do

1. If your school is made of brick, check for rounded corners, cracks, chips, or rough surfaces in the brick. Does the mortar stick out or recede? Does the brick look old? If your school is made of something other than brick, can you find evidence of weathering?
2. Check the foundation of your school. Are there any rough and irregular surfaces, cracks, or broken corners? Compare the foundation on each side of the building. Do you notice any difference in the amount of weathering?
3. Find a rock. Break away a portion to expose a fresh surface. Compare the exposed and fresh surfaces.
4. Look for moss or lichen growing on exposed rocks. Gently peel away the lichen and feel the rock surface underneath. Compare this surface to that of a rock without moss or lichen.
5. Examine sidewalks, stone fences, or concrete pillars. Is the concrete cracking, breaking, or crumbling? Are there trees or other plants growing nearby whose root systems may be affecting the concrete in some way?

Name ______________________ Date __________ Class __________



**Keeping a Record**

Summarize your findings in the table below. An example is provided.

| Location | Evidence of weathering | Cause(s) | Comments |
| --- | --- | --- | --- |
| sidewalk | cracks, broken corners | frost, tree roots | built in 1985—only a few cracks |
| | | | |
| | | | |
| | | | |

Name ______________________ Date ____________ Class ____________

# EXPLORATION 4

**Chapter 25**
Exploration Worksheet

## Monuments to Change, page 483

| **Your goal** | to learn about the relationship between weathering and time |
|---|---|

If there is a graveyard in your community that is over 100 years old, you have the makings of an interesting project.

### Problem 1

What is the relationship between the amount of weathering and the length of time that a monument has been exposed? Hint: Find stones of different ages made from the same kind of rock. Look for newer ones (0–100 years old), older ones (over 150 years old), and some in between. Compare the amount of weathering in each. Keep track of your findings in your ScienceLog, and then summarize your findings here.

______________________________

______________________________

______________________________

### Problem 2

Do different types of stones weather at different rates? Hint: Find several gravestones of approximately the same age but made of different kinds of rock. Compare the amount of weathering in each.

Photo also on page 483 of your textbook

______________________________

______________________________

______________________________

______________________________

What do you conclude from your study?

______________________________

______________________________

______________________________

______________________________

______________________________

Name ______________________ Date ____________ Class ____________

# EXPLORATION 5

**Chapter 25**
Exploration Worksheet

## Three Simulations, page 484

page 484

| **Cooperative Learning Activity** | |
|---|---|
| **Group size** | 3 to 4 students |
| **Group goal** | to simulate factors that might affect the speed of weathering |
| **Individual responsibility** | Each member of your group should choose a role such as leader, materials manager, safety monitor, or recorder. |
| **Individual accountability** | Each group member should be able to express what he or she learned by performing the simulations. |

It is very difficult (not to mention very time consuming) to determine the speed of weathering through actual observation. Real rocks take a very long time to break down. Sometimes, however, laboratory activities can *simulate* (imitate) what happens in the environment. Three factors that might affect the speed of weathering are investigated in the simulations that follow.

### You Will Need

- a magnifying glass
- rock samples (granite, limestone, sandstone, brick, and a piece of concrete)
- a tray
- a freezer

### Simulation 1

What effect do freezing and thawing have on rocks? Make a prediction.

______________________________

______________________________

### What to Do

1. Examine your samples with the magnifying glass. Can you find any cracks where water might enter?
2. Soak each rock sample in water. After several minutes, remove it and place it on a tray in the freezer overnight.
3. The next day, remove the samples from the freezer, and allow the rocks to warm up to room temperature.
4. Examine the samples for any changes. Then repeat the process of soaking, freezing, and thawing. If time allows, repeat the freezing-thawing cycle several times.

### Questions

1. Compare what happens to the different samples. Which sample shows the greatest change? the least change?

______________________________

______________________________

2. How does this simulation relate to conditions in the environment?

________________________________________

________________________________________

________________________________________

________________________________________

## You Will Need

- a whole piece of chalkboard chalk
- dilute hydrochloric acid
- a graduated cylinder
- two 250 mL beakers
- latex gloves

## Simulation 2

### What to Do

1. In addition to your safety goggles, wear latex gloves and an apron during this activity. Break a piece of chalkboard chalk in half. Put aside one half, and break the other half into several small pieces. (You can wrap it in a paper towel and hit it gently with a hammer.)
2. Carefully pour 100 mL of dilute hydrochloric acid into each beaker.
3. Place the large piece of chalk in one beaker and the smaller pieces in the other.
4. Clean up your lab area and wash your hands when you are finished.

### Questions

1. In which beaker did the bubbling last the longest?

________________________________________

________________________________________

2. What happened to the chalk in each beaker?

________________________________________

________________________________________

________________________________________

________________________________________

________________________________________

________________________________________

3. What comparison can you make between this investigation and the natural weathering process?

## Simulation 3

Do some kinds of rocks resist weathering better than others? One class investigated this problem by using marble chips and quartzite. They took out 100 g of each rock and placed each in a container about half full of water. After putting on the cover, they shook each container for 5 minutes at a rate determined before the experiment. After draining each container, they weighed and recorded the mass of the chips and quartzite. They repeated the procedure three more times and found the mass of each sample after each 5-minute period of shaking, for a total of 20 minutes. Their results are recorded in the table below.

| Time (min.) | Mass of marble chips remaining (g) | Mass of quartzite chips remaining (g) |
|---|---|---|
| 0 | 100.0 | 100.0 |
| 5 | 98.5 | 99.7 |
| 10 | 96.9 | 99.6 |
| 15 | 95.3 | 99.3 |
| 20 | 92.7 | 99.2 |

## Questions

1. Which kind of rock lost the greatest amount of mass?

2. What inferences can you make about the resistance of marble to weathering? the resistance of quartzite to weathering?

3. During which time period did the greatest loss of mass from the marble chips occur? Can you explain this?

______

______

______

4. What do you think might happen if you shook each sample for 1 hour?

______

______

______

5. How does this investigation simulate the weathering process?

______

______

______

## How Fast Does Weathering Occur in Your Community?

Weathering occurs at different rates, depending on the conditions in the environment. Suggest conditions in the environment that affect the speed of weathering (other than those you have already dealt with, of course). Make a list of all the factors in your area that might affect the speed of weathering.

______

______

______

______

______

______

______

______

______

Name ______________________ Date __________ Class __________

# Exploration 6

Chapter 25
Exploration Worksheet

## A Model Landslide, page 489

### Cooperative Learning Activity

**Group size** 3 to 4 students

**Group goal** to investigate a different variable of the landslide model and to demonstrate the result to the rest of the class

**Individual responsibility** Each member of your group should choose a role such as facilitator, materials coordinator, reporter, or timekeeper.

**Individual accountability** Each group member should be able to list the variable that his or her group investigated and give a brief summary of the group's results and conclusions.

Each year one of the projects in Mrs. Jefferson's class is to build a model and do an experiment with it. Pam and Sanjay decided to build a model that would help them figure out what causes landslides. Pam and Sanjay thought of the following ways to get the sand to slide:

1. Pour water on the sand.
2. Pile the sand deeper.
3. Make the angle of the board steeper.

### You Will Need

- a shallow, rectangular pan
- sand
- water
- a piece of plywood
- wooden blocks

### What to Do

1. Test Pam and Sanjay's methods. (You can see their model on page 489 of your textbook.) What other methods might they have tried?

______________________

______________________

______________________

Do all these methods cause the sand to slide down the board?

______________________

______________________

2. Can you think of any natural occurrences that match your investigations?

______________________

______________________

______________________

______________________

______________________

______________________

Name ______________________________ Date ____________ Class ____________

**Chapter 25**
**Review Worksheet**

## Challenge Your Thinking, page 491

### 1. Artist's Choice

You are a sculptor, and you have been asked to design a monument that will be placed in downtown New York City. How would you go about picking the type of stone that would be best to use? What would you have to know?

______________________________________________

______________________________________________

______________________________________________

______________________________________________

______________________________________________

______________________________________________

______________________________________________

______________________________________________

______________________________________________

______________________________________________

______________________________________________

Illustration also on page 491 of your textbook

### 2. Physical or Chemical?

There are two types of weathering: physical and chemical. Limestone dissolving to form caves is an example of chemical weathering. The cracking of rock during a freeze-thaw cycle is an example of physical weathering. Classify the following examples of weathering as *chemical, physical,* or *not enough information to say.*

____________________ A rock splits open after a day in the hot sun.

____________________ A block of sandstone grows soft and crumbly after being exposed to water for a period of time.

____________________ A rock breaks into pieces when it is heated.

____________________ A rock develops a grayish color after being left outside.

____________________ A rock is ground down during a flash flood.

____________________ A rock's surface is crumbly where a lichen grew on it.

**3. Please Publish**

In your ScienceLog, write a human-interest magazine article entitled "I Survived the Frank Landslide" or "Buried Alive in an Avalanche!" Be sure to include a description of the cause of the disaster.

**4. Time Warp**

After a 20-year absence, Rip returns to his hometown. He is dumbfounded. "Why, it's just as I left it. It hasn't changed a bit. The old sycamore tree we used to have the swing on is still there by the old stone church. Look—the tree still has my initials scratched in it. It hasn't changed and neither has the church. The creek is still crystal clear. It hasn't changed . . . " Were things truly just as Rip left them, or should he take a closer look? What would Rip see if he looked more closely?

________________________________________

________________________________________

________________________________________

________________________________________

________________________________________

________________________________________

________________________________________

________________________________________

**5. Designer Display**

Design a bulletin-board display whose theme is geological changes. Ask yourself the following questions:

- Who is my audience?
- What information do I want to convey?
- How can I best convey the information—through articles? through photographs? through drawings?
- How can I make the display interactive enough to keep my audience interested?

**Chapter 25**
**Assessment**

## Word Usage

**1.** In addition to the fact that they all start with *w,* what do the words *wind, wear, water,* and *weathering* have in common?

______________________

______________________

______________________

______________________

## Correction/ Completion

**2.** The following sentences are incorrect or incomplete. Make the sentences correct and complete.

**a.** Erosion can be caused by rain, wind, glaciers, and other agents, but all forms of erosion involve ____________.

**b.** Trees and other plants can prevent weathering by breaking up rocks, and they can cause erosion by holding soil in place.

______________________

______________________

______________________

## Numerical Problem

**3.** Every 2000 years, about 6 cm of land is removed from the continents by erosion.

**a.** How much is removed in 1 million years? Show your work.

______________________

______________________

**b.** How much is removed in 3 million years? Show your work.

______________________

______________________

**c.** Why are geologists interested in this kind of data?

______________________

______________________

______________________

______________________

Name ______________________ Date ____________ Class ____________



CHALLENGE 1

**Short Essay**

4. Could certain human activities affect the rate of weathering in an area? Explain.

_______________________________________________

_______________________________________________

_______________________________________________

_______________________________________________

_______________________________________________

_______________________________________________

_______________________________________________

_______________________________________________

_______________________________________________

_______________________________________________

_______________________________________________

_______________________________________________

_______________________________________________

**Answering by Illustration**

5. Suppose that you were given the supplies to build a medium-sized home of your own design. Design a structure that will resist the effects of weathering and erosion. Label the weather-resistant features.

Name _______________ Date _______ Class _______

Chapter 26
Transparency Worksheet

# The Water Cycle Teacher's Notes

This worksheet corresponds to Transparency 86 in the Teaching Transparencies binder.

## Suggested Uses

Use as a visual aid with any of the topics listed below.

The Power of Water, page 494
Water Below! page 497
Unit 4, Changes of State
Unit 5, Chemical and Physical Changes

Use the transparency with the transparency worksheet on the next page for reteaching or review. Please note: The worksheet is a reproduction of the actual transparency with certain labels omitted. For answers to that worksheet, see the transparency.

## Possible Extension Questions

1. As a class, trace each step of the water cycle. Have students record their answers on copies of the transparency worksheet.
2. What is the water cycle?
3. What is runoff?
4. Does the shape of the land and the amount of rainfall affect runoff? Explain.

## Answers to Extension Questions

1. Compare student answers with the actual transparency.
2. The water cycle is the process of evaporation and condensation that purifies the Earth's water.
3. Runoff is excess water that cannot be absorbed by the ground and that collects and flows over the Earth' surface.
4. Yes. The steeper the slope, the greater the speed of runoff. The amount of runoff is increased by heavy rains because the amount of time does not allow rain to soak into the soil. If the soil is already saturated, the amount of runoff will be greater.

Name ______________________ Date ____________ Class ____________

**Chapter 26**
Transparency Worksheet

# The Water Cycle

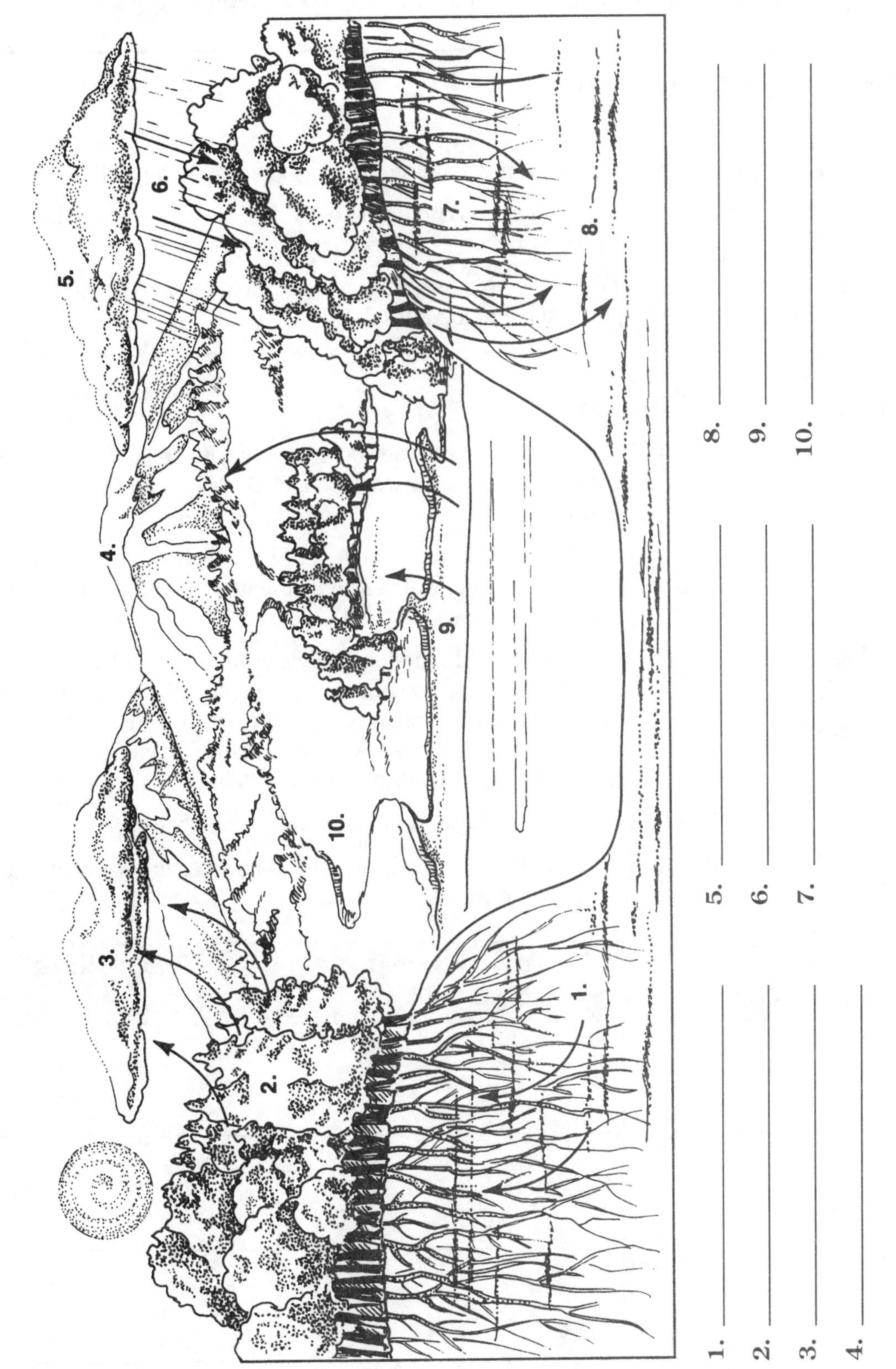


1. ____________________
2. ____________________
3. ____________________
4. ____________________
5. ____________________
6. ____________________
7. ____________________
8. ____________________
9. ____________________
10. ____________________

Chapter 26

# EXPLORATION 1

**Chapter 26**
Exploration Worksheet

## Investigating the Movement of Water in Soil, page 496

### Cooperative Learning Activity

**Group size** 3 to 4 students

**Group goal** to investigate the movement of water through different types of soil

**Individual responsibility** Each member of your group should choose a role such as primary investigator, investigative associate, supplies coordinator, or timer.

**Individual accountability** Each group member should be able to answer the questions asked in the final paragraphs of this Exploration.

### You Will Need

- three 500 mL beakers
- 3 soil samples
- water
- a graduated cylinder
- a watch with a second hand

### What to Do

1. Fill each container to the 400 mL mark with soil taken from several locations (sand, clay, garden soil, etc.). Be careful not to pack down the soil.
2. Look carefully at each sample. Predict which will hold the most water. Also predict which sample the water will move through the fastest.
3. Fill the graduated cylinder with water to the 100 mL mark. Slowly add as much of the 100 mL of water as possible to each soil sample until the soil sample can hold no more. Make note of how much water you used. At the same time, measure how long it takes for the water to reach the bottom of each container.
4. Record all of your observations in the table below.
5. When you have finished, clean up your area, and wash your hands with soap and water.

| Type of soil | Time for water to pass through | Volume of water absorbed |
|---|---|---|
| | | |
| | | |
| | | |

## Making Sense of Your Results: Porosity and Permeability

Certain soils allow water to pass through them faster than others do. In other words, some soils are more *permeable* than others. When you measure the time it takes for water to pass through a soil, you are measuring the **permeability** of the soil. Which sample was the most permeable? the least permeable?

______________________________________________

______________________________________________

Because of gravity, rainwater moves downward into tiny spaces in soil and rock. These spaces are called pores. The greater the volume of water that a soil can hold, the more *porous* the soil is. The volume of water that a soil can hold is a measure of the soil's **porosity.** Which sample was the most porous? the least porous?

______________________________________________

______________________________________________

Which type of soil would cause the most runoff during a rainstorm? Explain your answer based on the concepts of porosity and permeability.

______________________________________________

______________________________________________

______________________________________________

______________________________________________

______________________________________________

______________________________________________

______________________________________________

______________________________________________

Name ______________________ Date ____________ Class ____________

# EXPLORATION 2

Chapter 26
Exploration Worksheet

## Fast or Slow Streams, page 500

| Cooperative Learning Activity | | Safety Alert! |
|---|---|---|
| **Group size** | 3 to 4 students |  |
| **Group goal** | to investigate the movement of water through different types of soil | |
| **Individual responsibility** | Each member of your group should choose a role such as timer, materials manager, results tracker, or experiment coordinator. | |
| **Individual accountability** | Each group member should be able to answer the following question: What determines whether a stream flows quickly or slowly? | |

### You Will Need

- a setup such as the one shown on page 500 of your textbook (Be sure that there is a drain at the lower end of the pan.)
- a metric ruler
- a wax pencil
- food coloring
- a watch with a second hand

### Activity 1

What effect does changing the slope of a stream have on the speed of the water?

### What to Do

1. Write down a prediction for the question above.

   ______________________________________________

   ______________________________________________

2. Set up the pan of water and the siphon as shown on page 500 of your textbook. Adjust the trough so that the upper end is 4 cm high.
3. With a wax pencil, mark a starting line on the side of the trough.
4. Open the siphon. Add a drop of food coloring to the water as it flows past the starting line.
5. Time how long it takes the dye to reach the end of the trough.
6. Change the height of the trough to 8 cm. Then repeat steps 4 and 5. In your ScienceLog, compare the results.

## Activity 2

What effect does changing the volume of the water in a stream have on the speed of the water?

## What to Do

1. Write down a prediction for the question above.

______________________________________________

______________________________________________

2. Set up an experiment to answer this question. Consider the following as you plan what to do:
   - Can you use the same setup as before?
   - How can you increase the volume of water?
   - How will you determine the speed of the water?
   - Will you change the slope, as you did in Activity 1?
   - How will you determine whether the volume affects the speed of the water?
3. Can you think of another setup that might be used to answer this question?

______________________________________________

______________________________________________

## Analysis Please!

1. Look at the predictions you made. Does the data that you recorded support your predictions? In your ScienceLog, summarize your data, and present your conclusions.
2. You have identified two variables that affect how fast water flows in a trough. What other variables might affect the flow of a stream in a natural setting?

______________________________________________

______________________________________________

3. Which stream do you predict can carry the most sediment, a fast-flowing stream or a slow-flowing one? What evidence do you have to support your prediction? Explain why the speed of a stream affects the amount of sediment that it can carry.

______________________________________________

______________________________________________

______________________________________________

______________________________________________

______________________________________________

Name ____________________ Date __________ Class __________

# EXPLORATION 3

**Chapter 26**
Exploration Worksheet

## A Stream-Table Experiment, page 503

### Cooperative Learning Activity

| | |
|---|---|
| **Group size** | 3 to 4 students |
| **Group goal** | to conduct an experiment and to gather data on stream formation |
| **Individual responsibility** | Each member of your group should choose a role such as primary investigator, supportive assistant, materials manager, or recorder. |
| **Individual accountability** | Each group member should be able to answer the questions posed on page 502 of his or her textbook and to answer the question asked at the end of this Exploration. |

### Safety Alert!

### You Will Need

- a setup as shown in the photograph on page 503 of your textbook
- sand
- a metric ruler
- food coloring

### What to Do

1. Place sand to a depth of 5 cm in the stream table.
2. Wet the sand so that it can be shaped. Form the sand into a gentle slope.
3. Trace an S-shaped path with your finger almost to the bottom of the sand.
4. Turn on the water and adjust the flow so that it moves slowly into the tray.
5. Allow the water to flow down the path for 20 minutes. While the water is flowing, note the following locations:
   - **a.** places where the water slows down or speeds up (food coloring may help here)
   - **b.** places where banks are being eroded
   - **c.** places where sand buildup occurs
   - **d.** changes in the course of the stream
6. Turn off the water, but leave the sand in place.

Pretend that a friend was absent from school during the stream-table experiment. In your ScienceLog, describe to your friend exactly what you saw during the experiment. Do your observations support your choice of lot?

Name ______________________ Date __________ Class __________

Chapter 26
Exploration Worksheet

# EXPLORATION 4

## Check It Out, page 505

| **Your goal** | to examine the effects of rivers |
|---|---|

1. Is there a river near you? What stage is it in?

______________________________

______________________________

2. Many states contain well-known valleys or canyons. Find out about features such as these in your state. Then use a map to find out whether or not rivers flow through them. Try to answer these questions:

   **a.** Examine the photograph of the Grand Canyon on page 505 of your textbook. Which came first, the canyon (or valley) or the river?

   ______________________________

   ______________________________

   ______________________________

   ______________________________

   ______________________________

   **b.** What is the relationship between a valley, a canyon, and a river?

   ______________________________

   ______________________________

   ______________________________

   ______________________________

   ______________________________

   **c.** A river may have flowed through a valley or canyon at one time, even though there isn't a river flowing there now. How can you tell?

   ______________________________

   ______________________________

   ______________________________

   ______________________________

   ______________________________

Name ______________________ Date ____________ Class ____________

**Chapter 26**
**Math Practice Worksheet**

# The Grand Canyon Versus the Mighty River

Do this activity after reading Deltas and Estuaries, on page 506 of your textbook.

---

Imagine that you are a geologist and that you come across the following excerpt from a geological journal:

> **EARTH ALERT:** A gradual change in the global climate is causing the Colorado River to slowly deposit sediment in the Grand Canyon. Scientists estimate that the present rate of deposition is raising the canyon floor by 0.05 mm per year.

Your geological interests lead you to ask some questions. Suppose that you've organized your questions and concerns into the following itemized list. Using your mathematical knowledge, answer the following five items about the fate of the Grand Canyon.

---

**Item 1**

If the canyon is 1500 m deep, how long will it be until the river completely fills the canyon with sediment? Show your work.

______________________________________________

______________________________________________

______________________________________________

______________________________________________

---

**Item 2**

In the space below, graph the amount of sediment deposited in the Grand Canyon over an average human lifetime (about 75 years in the United States). Make sure to label the *x*- and *y*-axes and to give your graph a title.

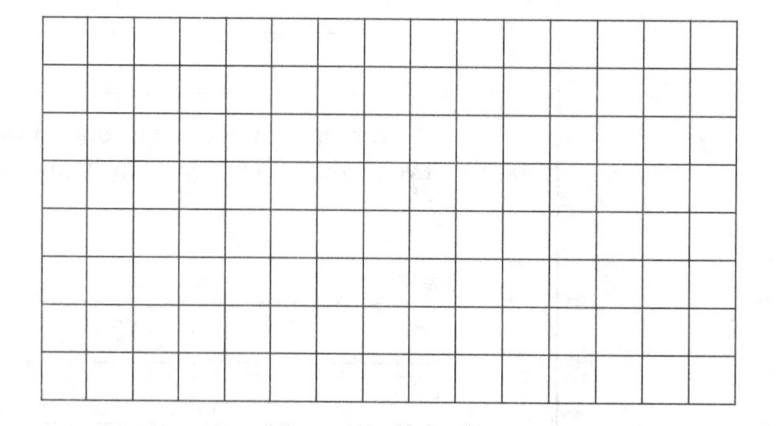

Name ______________________ Date ____________ Class ____________



**Item 3**

An animal is buried by the sediment at the bottom of the canyon. If a fossil hunter finds the animal 1 million years after the canyon is completely filled, how old is the fossil? Explain your answer.

**Item 4**

What will be the total volume of sediment dumped into the canyon if the canyon is 2 km wide and 30 km long? Show your work.

**Item 5**

If the same amount of sediment from Item 4 were instead carried to the mouth of the river and dumped into the ocean, it would form a delta. Assuming that the ocean is 500 m deep at the mouth of the river and that the elevation of the new land at the delta is at sea level, what will be the surface area of the delta? (Hint: volume/depth = area) Show your work.

## Glaciers Teacher's Notes

This worksheet corresponds to Transparency 88 in the Teaching Transparencies binder.

### Suggested Uses

Use as a visual aid with any of the topics listed below.

Glaciers—Rivers of Ice, page 512
Glaciers on the Move, page 514
Glacial Calling Cards, page 517
Glacial Landforms, page 518

Use the transparency to discuss the difference between alpine and continental glaciers.

As a class activity, have students prepare definitions for all of the terms listed on the transparency.

Use the transparency with the transparency worksheet on the next page for reteaching or review. Please note: The worksheet is a reproduction of the actual transparency with certain labels omitted. For answers to the worksheet, see the transparency.

### Possible Extension Questions

1. What is a glacier?
2. What causes a glacier to form?
3. Name two areas with large continental glaciers. Note: Students may not be able to answer the question completely, thus giving you a chance to extend the study of glaciers further.

### Answers to Extension Questions

1. A glacier is a large mass of ice and snow. Typically, a glacier moves slowly from the place where it accumulates to the point where it breaks up or melts.
2. A glacier is formed by snowfall in excess of the amount that melts over a long period of time.
3. Antarctica and Greenland

Name ______________________ Date ____________ Class ____________

**Chapter 26**
Transparency Worksheet

# Glaciers

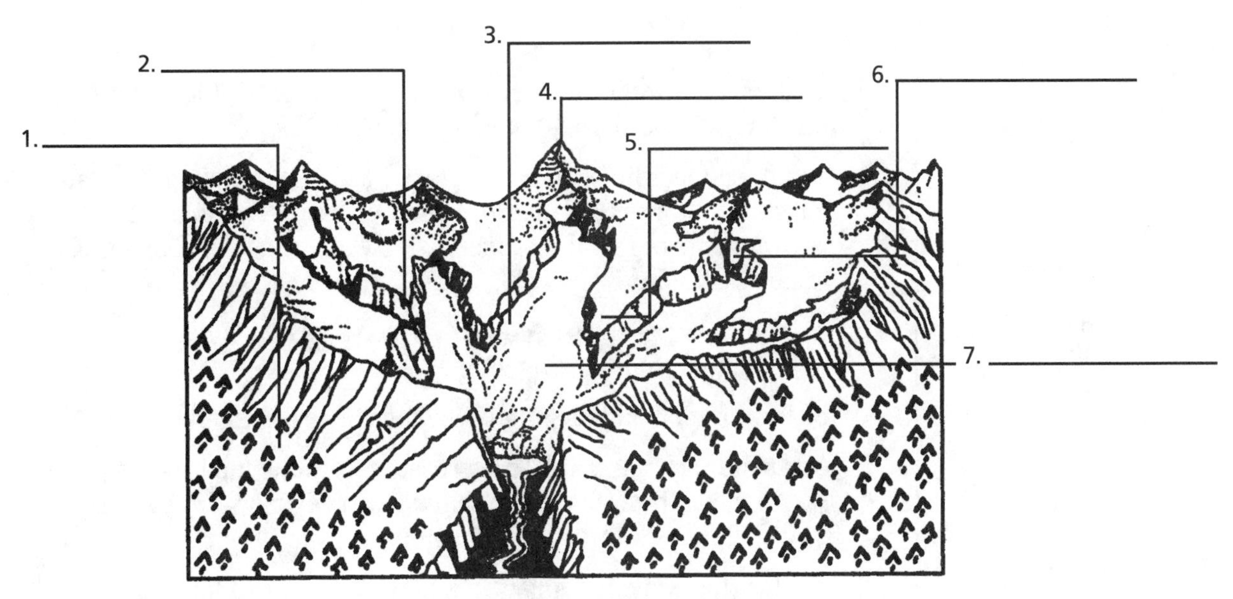


**Alpine Glacier**

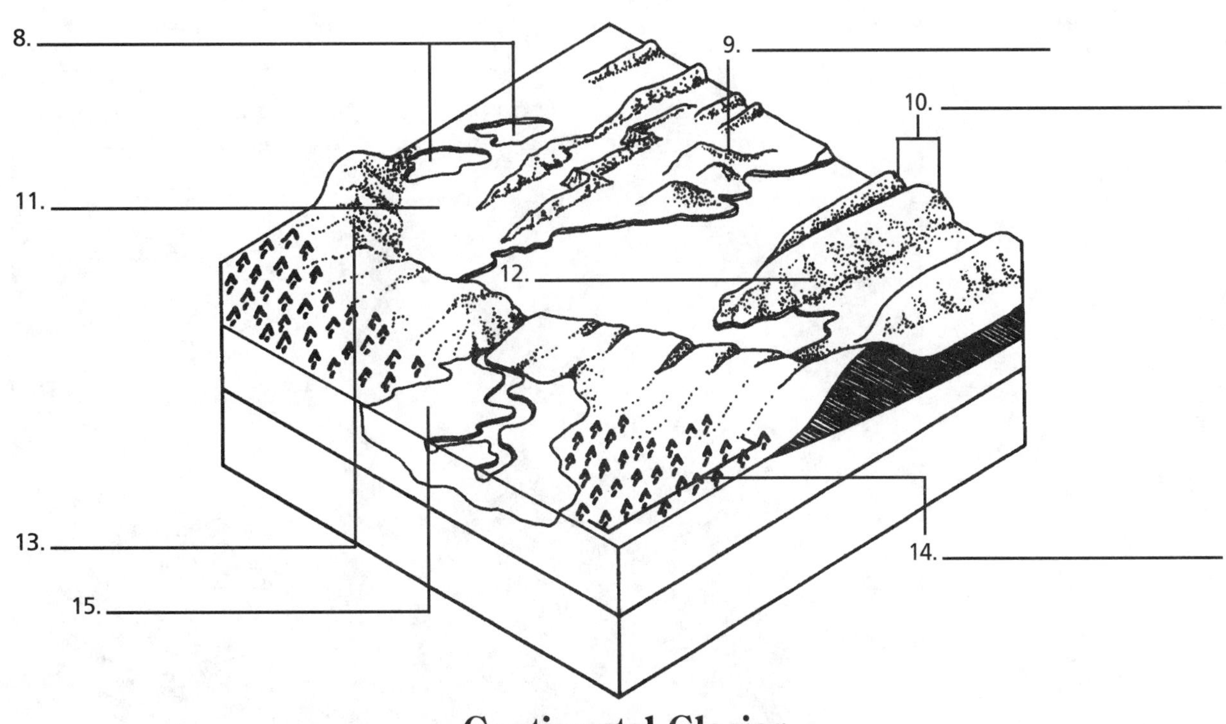


**Continental Glacier**

Name ______________________ Date ____________ Class ____________

# EXPLORATION 5

Chapter 26
Exploration Worksheet

## How Do Glaciers Form? page 513

| Your goal | to learn how snow turns into ice |
|---|---|

0 days

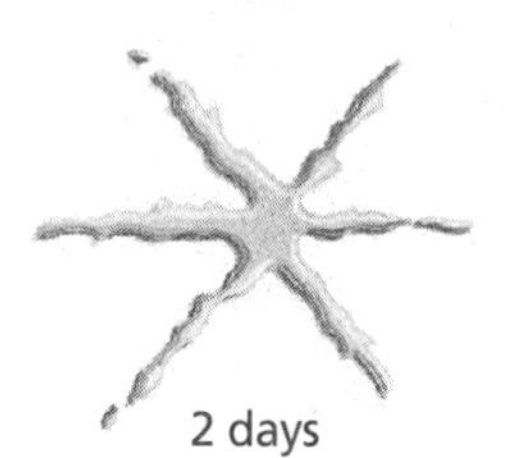

2 days

1 year

2 years

5 years

Illustration also on page 513 of your textbook

1. Can you turn snow into ice? How would you do it? It happens all the time in glaciers, although it takes quite a while.
2. To understand the process by which glaciers form, read the sentences below. Beware, though—their order is scrambled! Try to rearrange them into the proper order. Use the accompanying diagram as a guide.
   a. The snow particles become rounded pellets due to the added weight of accumulating snow.
   b. Evaporation and melting cause the flakes to lose their shape.
   c. As the snow builds up, the pressure from the top layers squeezes out the air between the rounded pellets of snow. Icy particles called *névé* (NAY vay) are formed.
   d. Star-shaped flakes collect in a fluffy mass with a lot of air caught between them.
   e. The névé undergoes further packing and melting until it gradually changes into solid ice.

   ______________________________________

3. Did you get the sentences in the right order? If you did, then you can truthfully say that you have got glaciers down cold!

Photo also on page 512 of your textbook

Name ______________________________ Date ____________ Class ____________

Chapter 26
Activity Worksheet

# Glaciers

Perform this activity after completing Exploration 5, How Do Glaciers Form? on page 513 of your textbook.

---

The illustrations below represent the changes in the size of a glacier over 50 years. Use the diagrams to answer the questions below.

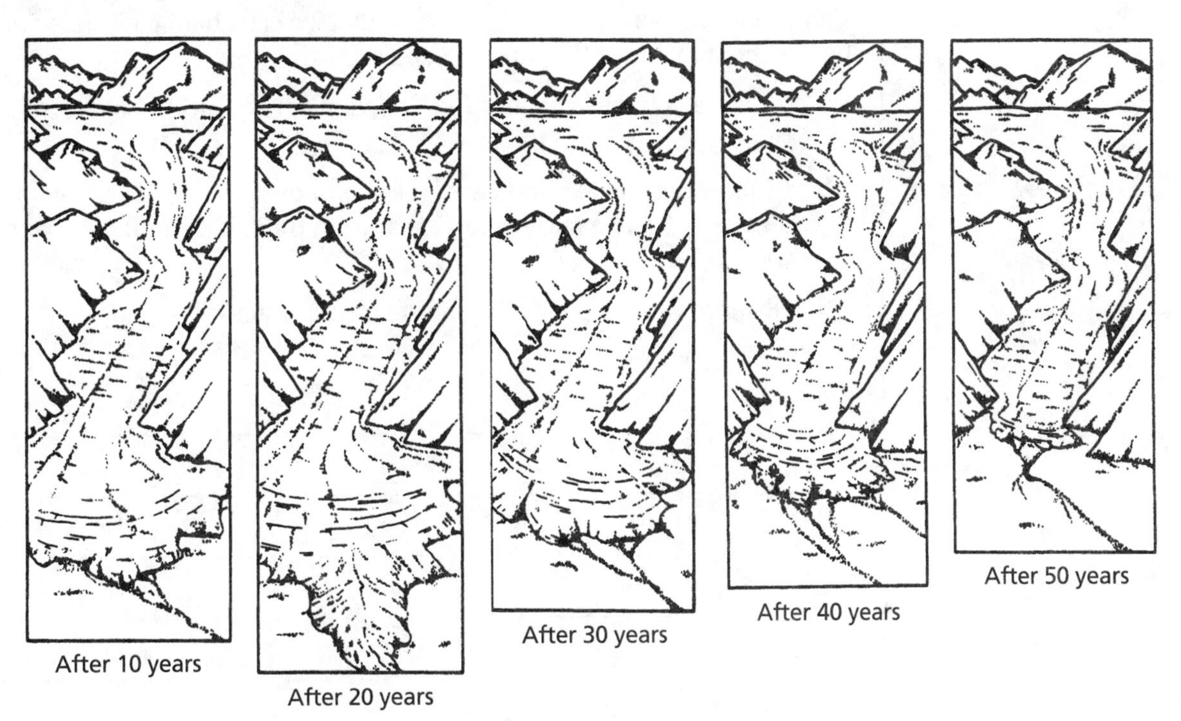


1. What happened to the glacier from 10 to 20 years?

_______________________________________________

2. What might have caused this change?

_______________________________________________

_______________________________________________

_______________________________________________

3. What happened to the glacier from 20 to 50 years?

_______________________________________________

4. What might have caused this change?

_______________________________________________

_______________________________________________

_______________________________________________

Name ____________________ Date ____________ Class ____________



# EXPLORATION 6

## Can Ice Really Flow? page 514

| **Your goal** | to determine whether ice can flow |
|---|---|

Can you design an experiment to prove that ice can change its shape or flow without melting? Perhaps you could try this idea.

1. Freeze a layer of ice about 1 cm thick in a tray. Once it has frozen, remove the ice from the tray.
2. In a freezer, support the layer of ice as shown on page 514 of your textbook. Hang a fairly heavy mass from the middle of the ice with a wire.
3. Observe what happens during the next few days.
4. Explain what you observe.

_______________________________________________

_______________________________________________

_______________________________________________

_______________________________________________

_______________________________________________

_______________________________________________

_______________________________________________

5. Repeat the experiment with different sized masses.
6. Can you apply your findings to explain how glaciers move?

_______________________________________________

_______________________________________________

_______________________________________________

_______________________________________________

_______________________________________________

_______________________________________________

_______________________________________________

Name ______________________ Date __________ Class __________

# EXPLORATION 7

**Chapter 26**
Exploration Worksheet

## Glaciers in Your Neighborhood, page 518

| **Your goal** | to determine whether your area was once covered by glaciers |
|---|---|

Look at the map below. Do you live in an area that was once covered by glaciers? If so, then you can probably find evidence of glaciation in your area. Organize a field trip to try to find that evidence.

Depending on your area, you may be able to find several different types of glacial features. Consult your teacher, reference materials, or a geologist to find out what types of glacial features, if any, you are likely to find.

Next, in your ScienceLog, prepare a field guide to use on your trip. It should include descriptions of typical evidence of past glacial activity.

Then, after studying a site, make complete notes and several sketches. Finally, write a field report in your ScienceLog to sum up your findings.

Illustration also on page 519 of your textbook

Chapter 26

Name ______________________ Date ____________ Class ____________

**Chapter 26**
Activity Worksheet

## Water Words

Try this activity before moving on to Challenge Your Thinking, on page 520 of your textbook.

---

Find each of the words listed here in the word-search puzzle below. The words can be found by searching up, down, across, and diagonally. When you find a word in the puzzle, circle it.

| | | | |
|---|---|---|---|
| delta | iceberg | porosity | snow |
| esker | meltwater | rain | stream |
| estuary | mouth | river | *Titanic* |
| fjord | outwash | runoff | tributary |
| glacier | permeability | sediment | watershed |

| | | | | | | | | | | | | | | |
|---|---|---|---|---|---|---|---|---|---|---|---|---|---|---|
| B | I | H | Q | I | Q | V | F | G | T | X | Y | I | O | Y |
| N | T | P | E | R | M | E | A | B | I | L | I | T | Y | F |
| H | N | Q | X | B | H | O | E | P | M | X | M | E | R | Y |
| B | E | Q | D | Y | Q | V | I | X | M | A | S | R | A | G |
| X | M | R | U | N | O | F | F | D | E | T | R | E | T | R |
| I | I | I | H | N | X | C | N | R | U | P | E | K | U | E |
| R | D | V | Q | E | N | H | T | A | L | O | I | S | B | B |
| E | E | R | A | I | N | S | R | Y | C | R | C | E | I | E |
| T | S | M | Q | Y | X | Y | D | V | I | O | A | M | R | C |
| A | X | K | O | M | M | J | W | O | N | S | L | O | T | I |
| W | Y | P | L | H | V | M | K | R | A | I | G | U | D | U |
| T | R | O | U | T | W | A | S | H | T | T | L | T | T | K |
| L | D | E | L | T | A | S | T | N | I | Y | H | H | D | G |
| E | N | Z | D | E | H | S | R | E | T | A | W | K | X | L |
| M | R | I | V | E | R | F | J | O | R | D | Q | B | U | S |

Name ______________________ Date ____________ Class ____________

**Chapter 26**
Review Worksheet

## Challenge Your Thinking, page 520

### 1. Take a Hike!

If you followed a river upstream, you would notice that the river forks again and again. However, rivers almost never fork as they flow downstream. Why is this?



______________________________________

______________________________________

______________________________________

______________________________________

______________________________________

______________________________________

______________________________________

### 2. How Good a Geologist Are You?

To find out, take the following test. A list of observations is given below. Give one or more possible explanations for each one. Check your answers with your teacher to find out how you did.

| | |
|---|---|
| Super Geologist | 6 or 7 |
| Geologist | 4 or 5 |
| Future Geologist | 3 or less |

**a.** Deep scratches are discovered in the surface of highly polished bedrock.

______________________________________

______________________________________

______________________________________

**b.** A perfectly preserved body is found in ice several kilometers from where it disappeared 75 years before.

______________________________________

______________________________________

______________________________________

**c.** Over a period of 5 years, a beach disappears.

______________________________________

______________________________________

______________________________________

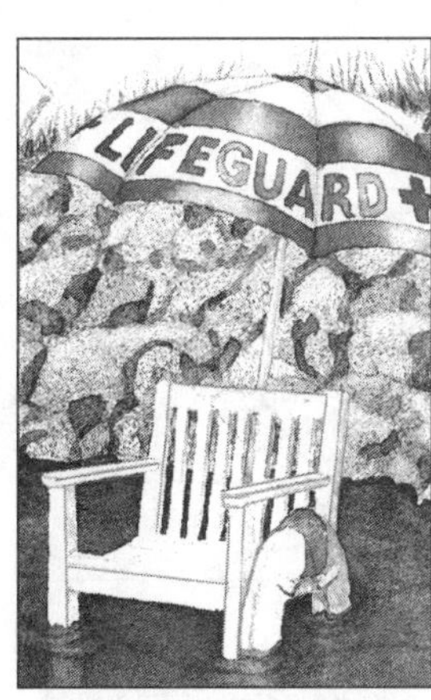


Illustrations also on page 520 of your textbook

**d.** A crack large enough to put your foot into runs the length of a huge boulder lying next to the highway.

______________________________

______________________________

______________________________

**e.** River water is salty many kilometers inland from the ocean.

______________________________

______________________________

______________________________

**f.** A huge granite boulder stands alone at the edge of a lake.

______________________________

______________________________

______________________________

**g.** A well goes dry in mid-August.

______________________________

______________________________

______________________________

## 3. Threats to Ground Water

Because of your research skills, a journalist has hired you to do background research for a magazine article. Find out why the ground water in parts of the United States may be unsafe to drink. As you do your research, write down information that the journalist could use in the article.

______________________________

______________________________

______________________________

______________________________

______________________________

______________________________

______________________________

______________________________

______________________________

**4. Wasting Away?**

If streams and rivers are such effective agents of erosion, why have they not reduced the entire United States to sea level?

___

___

___

___

___

___

Illustration also on page 521 of your textbook

**5. A Heated Debate Over Ice**

Sonia and Sunjay were traveling with their parents in the mountains. During their trip they visited a glacier and spent some time walking around on it. An argument broke out when Sonia said, "This glacier is moving, you know." "No way," replied Sunjay, "I can't see or feel it move." Who won the argument? Why?

___

___

___

___

___

___

___

Name ______________________ Date ____________ Class ____________

**Chapter 26**
**Assessment**

## Word Usage

1. Using what you know about the movement of rivers across land, put the following terms in sequential order: *mouth, delta, river,* and *ocean.* Then write one or two sentences describing your sequence.

______________________________________________

______________________________________________

______________________________________________

______________________________________________

______________________________________________

______________________________________________

______________________________________________

## Correction/ Completion

2. The following sentences are incorrect or incomplete. Your challenge is to make them correct and complete.

   **a.** Soil that allows large amounts of water to pass through it has a high porosity.

   ______________________________________________

   ______________________________________________

   ______________________________________________

   **b.** The upper surface of the ground-water zone is known as the

   ______________________.

## Short Response

3. Put the sentences below in order by numbering them from 1 to 6.

   ______ **a.** The gullies run together and form channels.

   ______ **b.** The river carries more water and begins eroding its valley as well as its bed.

   ______ **c.** A huge block of flat land is slowly lifted above sea level.

   ______ **d.** Tributaries cut downward to form V-shaped valleys.

   ______ **e.** Rainwater carves out gullies or ditches.

   ______ **f.** The river meanders as it erodes sideways.

CHALLENGE 1

## Short Response

4. Give a possible explanation for each of the following observations:

a. After a heavy rain, the river is brown.

_______________________________________________

_______________________________________________

_______________________________________________

_______________________________________________

b. On a farm, there is a shallow well that runs dry if it has not rained for a few weeks, but there is a deep well that always has water in it.

_______________________________________________

_______________________________________________

_______________________________________________

_______________________________________________

c. Many of the rivers in Maine flow more swiftly than many of the rivers in Iowa.

_______________________________________________

_______________________________________________

_______________________________________________

_______________________________________________

CHALLENGE 2

## Graph for Interpretation

5. A large area surrounding a stream has been heavily developed with homes, parking lots, and roads. Examine the graph below, which shows the amount of runoff into the stream before the area was developed and after the area was developed. Then answer the questions on the next page.

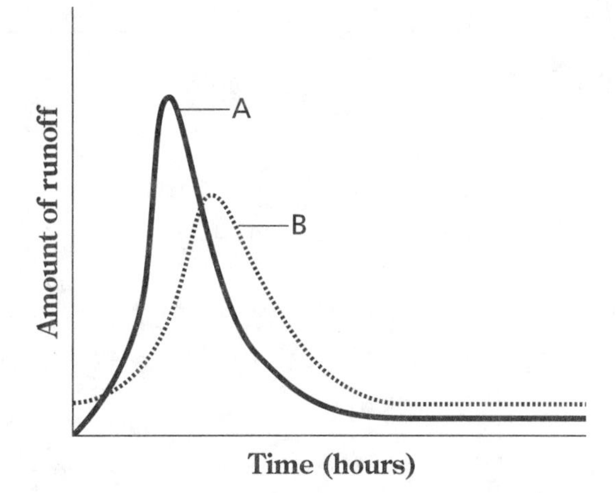

**a.** Which curve shows the amount of runoff after the area was developed? Explain.

**b.** Are floods more likely to occur before or after development? Explain.

Name ______________________ Date ____________ Class ____________

Unit 8
Unit Activity Worksheet

## Geology Match-Up

Match these geologic words and phrases as you conclude Unit 8.

Place a ruler on an imaginary line connecting each phrase on the left with the best match on the right. Each line will cross a letter and a number. The number tells you where to put the letter in the line of boxes at the bottom of the page. What is the message in the boxes?

|  |  |
|---|---|
| a natural change | outwash |
| any large movement of the Earth's crust | moraines |
| small rock fragments dropped by a glacier | downslope movements |
| glacial evidence | weathering |
| a crack in the sidewalk | tectonic movements |
| coastlines are shaped by these | weathered tombstone |
| avalanches and mudslides are called this | old rivers |
| excessive runoff may cause this | tides |
| low porosity | clay |
| rivers that meander | theory of continental drift |
| breakwaters prevent this | delta |
| flooded coastal valley | river channels |
| Alfred Wegener | glacial movement |
| fan-shaped wedge of sediment | estuary |
| gravity acting on ice | sand |
| has high permeability | flooding |
| gullies run together to form these | erosion |

9 C 16 Y 2 1 O K 8 R I 4 10 S F 14 11 P 6 O R U 13 7 3 R T 17 E 5 E 12 W 15

| 1 | 2 | 3 | 4 | | | 5 | 6 | 7 | 8 | | | 9 | 10 | | | 11 | 12 | 13 | 14 | 15 | 16 | 17 |
|---|---|---|---|---|---|---|---|---|---|---|---|---|---|---|---|---|---|---|---|---|---|---|
| | | | | | | | | | | | | | | | | | | | | | | |

Name ______________________ Date ____________ Class ____________

**Unit 8**
**Unit Review Worksheet**

## Making Connections, page 522

### The Big Ideas

In your ScienceLog, write a summary of this unit, using the following questions as a guide:

1. What is geology all about? (Ch. 24)
2. What does the expression "the present is the key to the past" mean? (Ch. 24)
3. How do geologists use inferences and observations? (Ch. 24)
4. What is some evidence that supports the theory of continental drift? (Ch. 24)
5. What are the main features of the Earth's interior? (Ch. 24)
6. Use your knowledge of the structure of the Earth to explain why the floor of the Atlantic Ocean is growing wider. (Ch. 24)
7. What are some of the agents of weathering? (Ch. 25)
8. What role does water play in the weathering process? (Ch. 25)
9. What role does gravity play? (Ch. 25)
10. What is some evidence for widespread glaciation in the past? (Ch. 26)

### Checking Your Understanding

1. A famous saying about geology is written in code below. Here is how to crack the code. One letter simply stands for another.

   C H K O O K L P A N E O

   G L O S S O P T E R I S

   Use the example to get started. Can you decipher this famous saying?

   P D A L N A O A J P E O P D A G A U

   ______ ______________ ____ ________ ________

   P K P D A L W O P — F W I A O D Q P P K J

   ____ ______ ________ — __________ ____________

2. Imagine that all tectonic activity suddenly came to a grinding halt. What would the Earth be like in 10 million years? 100 million years? 1 billion years? Continue your answer in your ScienceLog if necessary.

3. Most words have more than one meaning. Read each sentence below and explain the meaning of the underlined term. Then write a sentence to show that you understand the geological or scientific meaning of the underlined term as it was used in this unit.

**Example:** Betty and Sergei were in the runoff race.

Heavy rains can cause great damage due to runoff.

a. He is in big trouble for having his hand in the till.

b. How are you weathering the storm?

c. Why not meander over to my house after supper?

Name ______________________ Date ____________ Class ____________



d. Her election win was a landslide victory.

______________________________________________

______________________________________________

______________________________________________

______________________________________________

e. She received an avalanche of phone calls after she won the trophy.

______________________________________________

______________________________________________

______________________________________________

______________________________________________

f. The baseball missed home plate by a country mile.

______________________________________________

______________________________________________

______________________________________________

______________________________________________

Illustrations also on page 523 of your textbook

g. Is a water table like a water bed?

______________________________________________

______________________________________________

______________________________________________

______________________________________________

Find other examples of words used in this unit that have both a scientific and a common meaning, and record them in your ScienceLog.

4. How do you think the process of erosion would be affected if

a. the force of gravity were doubled?

______________________________________________

______________________________________________

______________________________________________

______________________________________________

**b.** the average global temperature went up by 5°C?

**c.** the atmosphere vanished?

**d.** the boiling point of water suddenly changed to 10°C?

**e.** there were no gravity?

**f.** water froze at 25°C?

g. water were four times heavier (for a given volume)?

______________________

______________________

______________________

______________________

5. concept map Make a concept map using the following terms or phrases: *waves, ice, gravity, erosion, wind,* and *flowing water.*

Illustration also on page 523 of your textbook

Name ______________________ Date __________ Class __________

**Unit 8**
**End-of-Unit Assessment**

## Word Usage

**1.** Compare the *porosity* and *permeability* of sand and garden soil. Write a sentence to show that you understand these terms.

______________________________

______________________________

______________________________

**2.** For each pair of words below, show the connection between the words by using them in a sentence.

**a.** *water table* and *lakes*

______________________________

______________________________

**b.** *deposition* and *deltas*

______________________________

______________________________

**c.** *continental drift* and *fossils*

______________________________

______________________________

## Correction/ Completion

**3.** The following statement is either incorrect or incomplete. Your job is to make it correct and complete.

During an ice age, ______________ advance over the surface of the Earth. When they stop, a ridge of deposit called a ______________ ______________ is formed.

## Short Responses

**4.** Read the observations and inferences below. Decide if each inference is reasonable. If not, write a reasonable inference on the next page.

| Observation | Harold's inference |
|---|---|
| **a.** A ball lies by a broken window. | The ball caused the window to break. |
| **b.** The Mississippi River floods in the spring. | Melting snow and spring rains cause the floods. |
| **c.** A river has carved a narrow channel through a rugged mountain valley. | The river dates back to the beginning of time. |

**d.** Granite boulders and rocks are found in farmland 200 km from mountains made of granite.

These boulders and rocks have always been there.

______________________________________________

______________________________________________

______________________________________________

______________________________________________

______________________________________________

**5.** Geologists use their observations to make inferences about what must have taken place in the past. Use your "geo-logic" to draw inferences about each of the following situations:

**a.** Some of the rock layers in a cliff contain fossils of fish, oysters, and clams.

______________________________________________

______________________________________________

______________________________________________

**b.** You find an exposed layer of rock whose surface bears deep grooves and appears to have been polished.

______________________________________________

______________________________________________

______________________________________________

## Illustrations for Interpretation

**6.** Which of the figures below would show the fastest rate of weathering, and why?

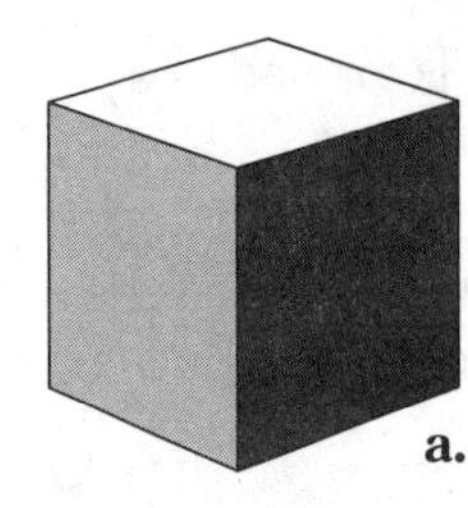

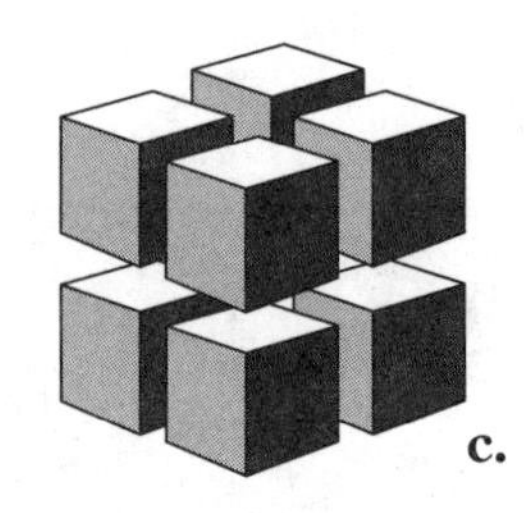


______________________________________________

______________________________________________

______________________________________________

______________________________________________

Name ______________________ Date __________ Class __________



7. Describe what is happening to the river in the diagram below.

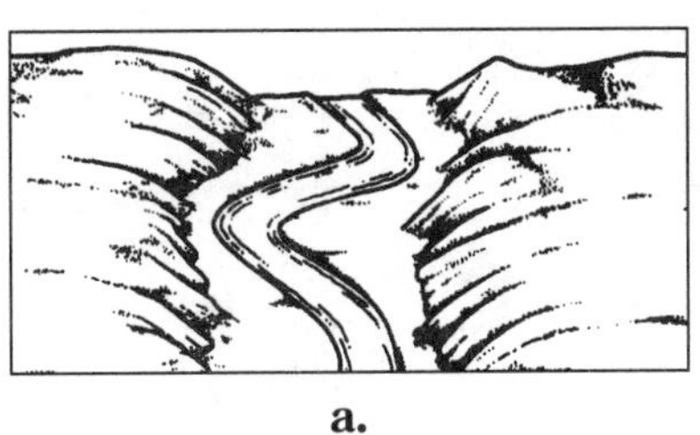

a.

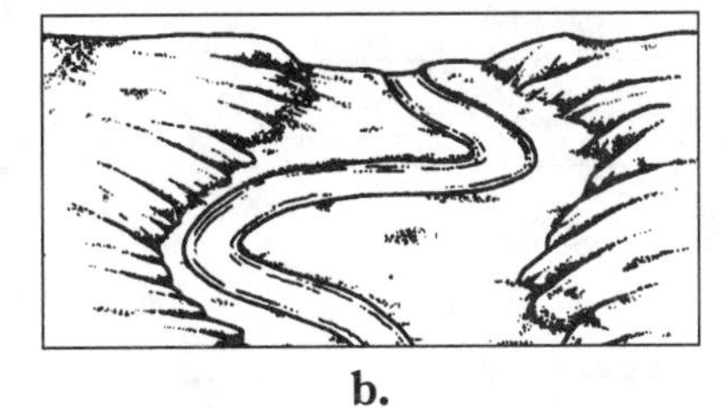

b.

c.

______________________________

______________________________

______________________________

______________________________

______________________________

**Illustration for Correction or Completion**

8. Sketch what you think the cliff shown below might look like after 100 years if the cliff were made of (a) hard granite, (b) soft sandstone, and (c) soil and clay. Include a written description with each sketch if you wish.

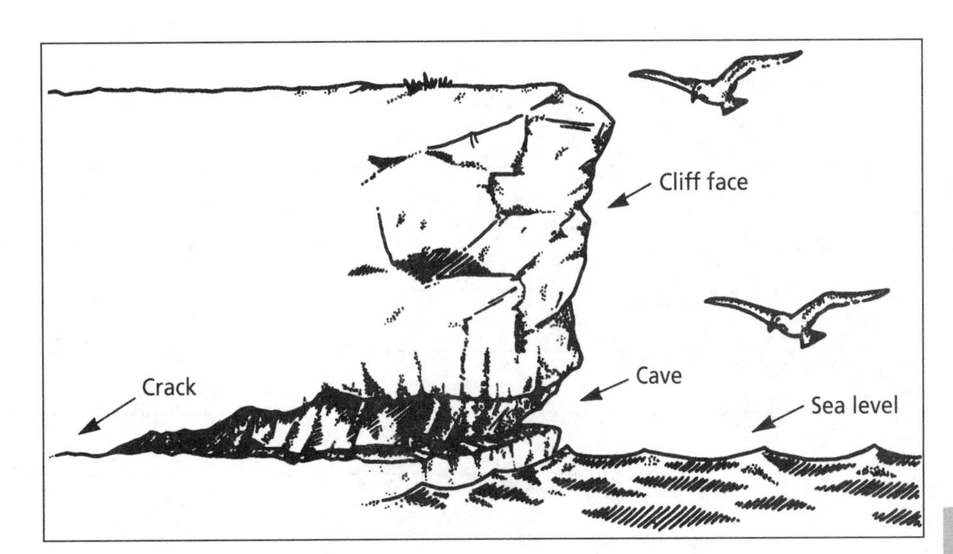

CHALLENGE 1

## Short Response

9. How is a glacier like a bulldozer?

______________________________________________

______________________________________________

______________________________________________

______________________________________________

______________________________________________

______________________________________________

## Short Essay

10. Factors that affect how fast weathering and erosion occur differ from place to place. Name as many differences as you can between the speed and amount of weathering in the following pairs of locations:

**a.** a mountain slope in the Sierra Nevadas and a beach in Florida

______________________________________________

______________________________________________

______________________________________________

______________________________________________

______________________________________________

______________________________________________

______________________________________________

______________________________________________

______________________________________________

______________________________________________

______________________________________________

______________________________________________

______________________________________________

**b.** the moon and a ranch in Brazil

## Illustration for Interpretation

**11.** The sketch below represents a new house built next to a hill. Mark the positions on the diagram where the owner may have some cause for concern. Then explain why the owner should be concerned.

## Short Essay

**12.** Arthur's family decided to build a swimming pool in their backyard. Bulldozers came in late fall and excavated the hole, but soon the temperature fell below freezing, and construction stopped. This gave Arthur an idea for a project on weathering and erosion. Write a brief account of the investigation Arthur might have done during the fall, winter, and spring.

Name ______________________ Date __________ Class __________



# Erosion Evasion — Teacher's Notes

**Overview** Students analyze and compare how three different rock samples resist mechanical erosion in water.

## Materials
### (per activity station)

- water
- 3 plastic jars with lids
- a piece of limestone (2–3 pieces of chalk would suffice)*
- a piece of granite*
- a piece of soft sandstone or loosely cemented conglomerate*
- a watch or clock with a second hand
- a magnifying glass

**Each rock sample should have the same mass.*

## Preparation

Prior to the assessment, equip student activity stations with the materials needed for each experiment.

## Time Required

Each student should have 20 minutes at the activity station and 15 minutes to complete the Observation Chart.

## Performance

At the end of the assessment, students should turn in the following:

- a completed Observation Chart

## Evaluation

The following is a recommended breakdown for evaluation of this Activity Assessment:

- 30% scientific approach to the experiment
- 40% clarity and thoroughness of observations
- 30% logic and thoughtfulness of conclusions

Name ____________________ Date __________ Class __________

**Unit 8**
Activity Assessment

# Erosion Evasion

Nothing on Earth can permanently escape the effects of erosion and weathering—but some materials can resist these forces better than others. At your activity station are three kinds of rocks. You will subject these rocks to mechanical erosion similar to that which occurs in a river. Your challenge is to analyze how each of these rock samples resists mechanical erosion. Then rank your samples according to their erosion resistance.

## Before You Begin . . .

As you work through the tasks, keep in mind that your teacher will be observing:

- whether you use a scientific approach
- how complete and accurate your observations are
- how thoughtful and logical your conclusions are

Now you're ready to observe!

## Task 1

Describe the appearance of each of your rocks before beginning your investigation. Record all of your observations in the Observation Chart. Include a prediction about how much you think each sample will be affected by erosion.

## Task 2

Place sample 1 into one of the jars and fill the jar halfway with water. Tightly seal the lid. Now vigorously shake the jar for 1 minute. Observe the results, and record any changes that you observe in the rock sample and the water.

## Task 3

Repeat the procedure above for the other two rock samples. Use a clean jar for each sample.

## Task 4

Summarize the effect of mechanical erosion in water for each sample. Then describe the erosion resistance of each sample, listing them in order from that with the highest level of resistance to that with the lowest level of resistance.

## Observation Chart

| Sample | Description of sample before erosion and prediction | Description of sample, water, and jar after erosion | Erosion resistance |
|---|---|---|---|
| 1. | | | |
| 2. | | | |
| 3. | | | |

Name ______________________ Date ____________ Class ____________

**Unit 8**
Self-Evaluation

## Self-Evaluation of Achievement

The statements below include some of the things that may be learned when studying this unit. If I have put a check mark beside a statement, that means I can do what it says.

______ I can explain how observation and inference are used to figure out what happened in the past to cause the Earth to be what it is today. (Ch. 24)

______ I can distinguish between the theory of continental drift and the theory of plate tectonics. (Ch. 24)

______ I can give several examples of changes in the Earth that are brought about by natural causes. (Ch. 25)

______ I can explain the processes of weathering and erosion and give examples of each. (Ch. 25)

______ I can identify several factors that affect the speed of water in a stream. (Ch. 26)

______ I can predict where banks will be eroded and where sand buildup will occur in a meandering stream. (Ch. 26)

______ I can illustrate the different stages that all rivers undergo as they age and can describe the characteristics of each stage. (Ch. 26)

______ I can write a paragraph on glaciers describing the location of glaciers now and in the past, how they form, how they move, and the effect they have on the land. (Ch. 26)

I have also learned to ______________________________________________

______________________________________________________________

______________________________________________________________

______________________________________________________________

______________________________________________________________

I would like to know more about ______________________________________

______________________________________________________________

______________________________________________________________

______________________________________________________________

______________________________________________________________

Signature: ______________________

Name ____________________ Date __________ Class __________

Unit 8 SourceBook Activity Worksheet

SourceBook

# A Rock-Cycle Simulation

Complete this activity after reading pages S180–S181 of the SourceBook.

| Your goal | to learn about the different stages of the rock cycle by simulating the formation of sedimentary, igneous, and metamorphic rocks | Safety Alert! 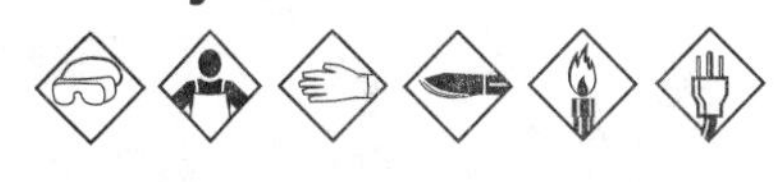 |
|---|---|---|

## You Will Need (per group)

- newspaper
- 2 to 3 each of red, green, blue, and yellow crayons
- a small pencil sharpener or scissors
- heavy-duty aluminum foil
- 2 large, heavy books, such as a dictionary or encyclopedia
- a watch with a second hand
- a metric ruler
- an aluminum pie plate
- oven mitts
- a hot plate

## What to Do

1. Form into groups of three or four students.
2. Cover your desktop or work area with two or three sheets of newspaper. Shave each crayon into fragments using a small pencil sharpener or a pair of scissors, keeping the crayon shavings separated into four piles according to color.
   **Caution: If using the scissors, be very careful not to cut yourself!**
3. What did the whole crayons represent in this rock-cycle simulation? What might the crayon shavings represent? the pencil sharpener or scissors?

   ______________________________

   ______________________________

   ______________________________

   ______________________________

   ______________________________
4. Tear off a sheet of aluminum foil about 20 cm × 20 cm. You may need to trim the sheet with scissors to get the proper dimensions. Fold this sheet in half. Place the foil on the newspaper, opening it so that you can see the crease in the middle.
5. Sprinkle the crayon shavings of one color in the center of the flat side of the foil so that they cover an area of approximately 4 cm × 4 cm to a depth of 0.5 cm. Then apply the other colors of crayon shavings, creating layers of color.
6. What might these layers of crayon shavings represent?

   ______________________________

   ______________________________

Name ________________ Date ________ Class ________



7. Fold the top half of the foil over the crayon layers. Place this foil package between two heavy books, and apply light pressure for 2 seconds. Remove the foil package from the books and open it.

8. What has occurred? What kind of rock did you simulate? What do the books represent in this simulation?

9. Place the crayon "rock" back in the foil, and put the foil between the two books again. This time both you and a partner press as hard as you can against the books for 1 minute. Remove the package from between the books and open it.

10. What happened to the crayon "rock layers" this time? What kind of rock has been simulated? How are the books used differently in this simulation?

11. Put the aluminum pie plate on the hot plate. Wrap the crayons in the foil again and place the package in the pie plate. Turn the hot plate on.

12. What do you think is happening to the crayon "rock"? **Caution: Do not touch the hot foil package!** What does the hot plate represent?

13. Turn off the hot plate and allow the crayon "rock" to cool and harden completely. What type of rock did you simulate?

14. How is this simulation process different from the rock cycle that actually occurs in nature?

# Unit CheckUp, page S205

## Concept Mapping

The concept map below illustrates major ideas in this unit. Complete the map by supplying the missing terms. Then extend your map by answering the additional question below.

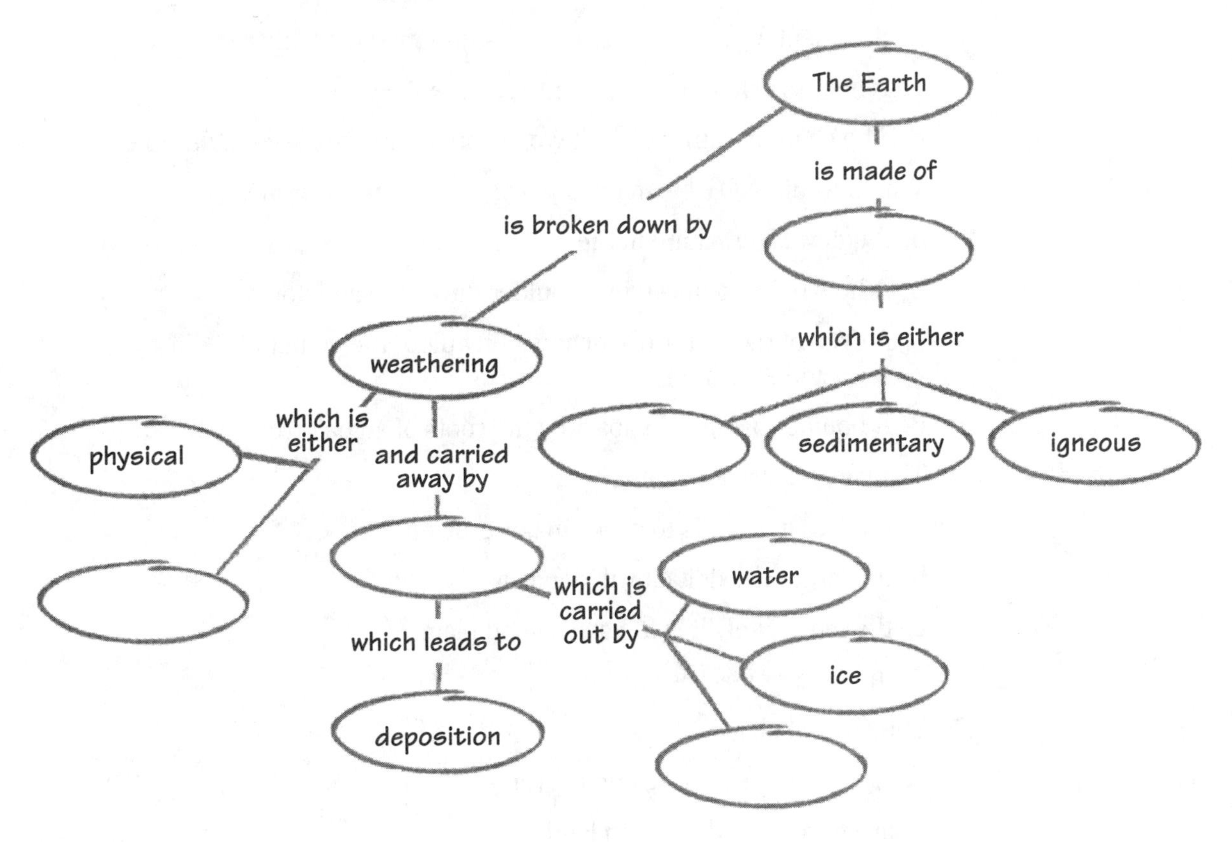


Where and how would you include the terms *rock cycle, gravity,* and *tectonic plates*?

Name ______________________ Date __________ Class __________



## Checking Your Understanding

Select the choice that most completely and correctly answers each of the following questions.

1. Which type of rock is formed from molten material?
   **a.** igneous  **b.** sedimentary
   **c.** metamorphic  **d.** magma
2. Which is true of the rock cycle?
   **a.** Any one rock type can be changed into any other.
   **b.** Sedimentary rock always changes into metamorphic rock.
   **c.** Igneous rock always turns into sedimentary rock.
   **d.** The change from one rock type to another takes several decades.
3. Which would NOT be an example of physical weathering?
   **a.** A sidewalk cracking in the summer sun
   **b.** A hole being scoured in a boulder during a sandstorm
   **c.** A drop of rainwater dissolving a tiny bit of the cement holding sandstone together
   **d.** A boulder being split apart by the roots of an oak tree
4. The faster a stream flows,
   **a.** the less likely it is to cause major erosion.
   **b.** the larger the delta it will deposit.
   **c.** the more likely it is to have a flood plain.
   **d.** the greater its erosive ability.
5. The abyssal plains
   **a.** are often swept by strong currents.
   **b.** are never found far from land.
   **c.** are sites of little activity.
   **d.** are home to many plants and animals.

## Interpreting Photos

The photo on page S206 of your textbook shows an unusual land feature that has been formed by the action of erosion on its surface. Using what you have learned in this unit, which type of erosion—gravity, streams, glaciers, wind, or waves—might have been responsible for producing this landmark? Explain.

______________________________

______________________________

______________________________

______________________________

## Critical Thinking

Carefully consider the following questions, and write a response that indicates your understanding of science.

1. In what major ways do wind erosion and water erosion differ?

2. Eventually, the Earth may lose so much heat that its interior becomes solid. What effect might this have on the Earth's landforms?

3. Weathering does not occur evenly throughout the world. Which types of climates probably experience the fastest weathering? the slowest? Why?

Name _______________ Date _______ Class _______



4. Tests have shown that, as you move away from the center of the mid-ocean ridges, the rocks become steadily older. How does this evidence support the idea that the Earth's crust moves?

5. Lakes are usually short-lived features, geologically speaking. Explain why this is so, using the concepts of erosion and deposition.

**Portfolio Idea**

As you know, the Earth is estimated to be about 4.5 billion years old. This is a span of time most people find very difficult to grasp. Design a scenario appropriate for helping an eight-year-old child to understand this span of time, using the major geologic events shown on page S184 of your textbook. For example, you might say "If the Earth's history were compressed into one year, oceans began forming on February . . ." and so on. Don't use this example. Be imaginative, but be accurate. Record your scenario in your ScienceLog.

Name ______________________ Date ____________ Class ____________





---

**1.** The Earth's surface is constantly changing.

**a.** true **b.** false

**2.** Rocks both on the Earth's surface and under it can undergo weathering.

**a.** true **b.** false

**3.** Soil contains ______, which is the decaying remains of dead plants and animals.

**a.** clay **b.** humus **c.** gravel **d.** dust

**4.** Soil that is high in humus is fertile.

**a.** true **b.** false

**5.** Which of the following does not cause a change in the composition of rock that is being eroded?

**a.** physical weathering **b.** chemical weathering

**6.** If you find rocks in the bottom of a river bed that are smooth with almost no sharp edges, you are probably observing an example of what type of weathering?

**a.** exfoliation **b.** expansion **c.** abrasion **d.** frost action

**7.** Piles of broken rock pieces at the bottom of a cliff or steep slope are called

**a.** slump. **b.** creep. **c.** talus. **d.** channels.

**8.** The most common form of erosion due to gravity is

**a.** slump. **b.** creep. **c.** talus. **d.** channels.

**9.** The Grand Canyon is an example of

**a.** erosion by gravity. **b.** erosion by streams.

**c.** erosion by glaciers. **d.** erosion by wind.

**10.** The Great Lakes were formed by

**a.** valley glaciers. **b.** continental glaciers.

**11.** A flood plain contains very poor soil and is not suitable for farming.

**a.** true **b.** false

12. Deltas are formed by

    **a.** stream deposits. **b.** glacial deposits.

    **c.** wind deposits. **d.** ocean currents.

13. Barrier islands may be found along the Gulf Coast of the United States.

    **a.** true **b.** false

14. Match the processes on the left with the correct type of weathering on the right.

| | | |
|---|---|---|
| ______ | frost action | **a.** physical weathering |
| ______ | oxidation | **b.** chemical weathering |
| ______ | abrasion | |
| ______ | root action | |
| ______ | hydration | |
| ______ | dissolving by acids | |

15. Match the definition on the left with the correct agent of weathering on the right.

| | | |
|---|---|---|
| ______ | Rocks come in contact with other rocks. | **a.** rebounding |
| ______ | A chemical in the air reacts with water and forms carbonic acid. | **b.** abrasion |
| ______ | Rock that is buried underground becomes exposed by erosion and expands. | **c.** carbon dioxide |
| ______ | Rock that expands when heated and contracts when cooled is stressed and eventually cracks or breaks. | **d.** thermal expansion |

16. Match the methods of deposition on the right with their effects on the left.

| | | |
|---|---|---|
| ______ | loess | **a.** wind |
| ______ | delta | **b.** ocean waves and currents |
| ______ | moraines | **c.** river |
| ______ | barrier islands | **d.** glacier |

**17.** ____________________ is an end product of physical and chemical weathering that helps support plant life.

**18.** A gradual process that plays a major role in shaping the Earth's landforms by carving away pieces of rock and soil is called ____________________.

**19.** Arrange the following pieces of weathered rock in order from largest to smallest. All are found in soil.

**a.** silt **b.** sand **c.** gravel

____________________

**20.** Explain what happens in the process of physical weathering called *root action.*

**21.** If a river's main channel is wide and flat, would it more than likely be an old river or a young river?

**22.** How does the absence of plants in deserts affect erosion?

**23.** What two factors contribute to the effect that waves have on a shoreline?

**24.** In the 1930s, the midwestern part of the United States fell victim to the Dust Bowl. Using what you learned in this SourceBook unit, explain what probably happened during the 1930s in that area.

**25.** What causes some beaches to be sandy and others to be rocky?

Estimado padre/madre de familia,

En las próximas semanas, su hijo o hija examinará la historia de la Tierra. Analizará el modo cómo ha cambiado el paisaje de nuestro planeta y algunas de las fuerzas que han causado estos cambios. Investigará también los efectos que producen el agua, el viento y el hielo en las formaciones terrestres. Cuando los estudiantes hayan terminado la Unidad 8, deberán ser capaces de dar respuesta a las siguientes preguntas, para captar las "grandes ideas" de la unidad.

1. ¿De qué trata la geología? (Cap. 24)
2. ¿Qué significa la expresión "el presente es la clave del pasado"? (Cap. 24)
3. ¿Cómo usan los geólogos las deducciones y las observaciones? (Cap. 24)
4. ¿Qué pruebas hay para sostener la teoría del desplazamiento continental? (Cap. 24)
5. ¿Cuáles son los rasgos principales del interior de la Tierra? (Cap. 24)
6. Usa tu conocimiento de la estructura de la Tierra para explicar por qué el fondo del Océano Atlántico se está haciendo más ancho. (Cap. 24)
7. ¿Cuáles son algunos de los agentes de desgaste? (Cap. 25)
8. ¿Qué rol tiene el agua en el proceso de desgaste? (Cap. 25)
9. ¿Qué rol tiene la gravedad? (Cap. 25)
10. ¿Cuáles son algunas de las pruebas de la existencia de extensos glaciares en el pasado? (Cap. 26)

Si Ud. lo desea, puede hacer a su hijo o hija algunas de estas preguntas a medida que vamos avanzando en la unidad. Será para el estudiante un ejercicio valioso de comunicación de sus conocimientos. No se preocupe si las respuestas no suenan como algo aprendido de memoria en el libro de texto, porque no lo son. Con ayuda de su libro, los estudiantes van a descubrir las respuestas por sí mismos, ya individualmente, ya en grupos. Esto es por un lado el desafío y por otro el lado divertido de la ciencia.

Atentamente,

Los materiales que aparecen abajo van a ser usados en clase para varias exploraciones de ciencia de la Unidad 8. Su contribución de materiales va a ser muy apreciada. He marcado algunos de los materiales en la lista. Si usted los tiene y quiere donarlos, por favor mándelos a la escuela con su hijo o hija para el

______________________________________.

- ○ cubos
- ○ velas
- ○ cartón (hojas finas)
- ○ tubos de cartón
- ○ pinzas de ropa o ganchos pequeños
- ○ recipientes (idénticos y transparentes)
- ○ colorante comestible
- ○ tapas de frascos
- ○ frascos (grandes)
- ○ guantes tipo “latex”
- ○ revistas
- ○ lupas
- ○ marcadores
- ○ plastilina para modelar
- ○ periódicos
- ○ recipientes de horno (llanos, rectangulares)
- ○ toallas de papel
- ○ jarras
- ○ madera compensada, de más o menos 20 cm × 30 cm
- ○ cartón para posters
- ○ muestras de rocas (granito, piedra caliza, arenisca, ladrillo, trozos de cemento)
- ○ caños de goma
- ○ arena
- ○ muestras de tierra (arena, arcilla, de jardín)
- ○ bandejas (grandes, de plástico o metal)
- ○ lápices de cera
- ○ alambre
- ○ cortadores de alambre
- ○ bloques de madera (grandes)

Desde ya, le agradecemos su ayuda.

En la Unidad 8, Nuestra Tierra cambiante, vas a estudiar el modo cómo se han formado los diversos paisajes. Vas a examinar algunas de las fuerzas que han formado el terreno y a investigar los efectos que producen el agua, el viento y el hielo en la configuración del terreno. Al ir leyendo la unidad, trata de responder a las siguientes preguntas. Estas son "las grandes ideas" de la unidad. Cuando puedas contestar estas preguntas, habrás logrado entender bien los principales conceptos de esta unidad.

1. ¿De qué trata la geología? (Cap. 24)
2. ¿Qué significa la expresión "el presente es la clave del pasado"? (Cap. 24)
3. ¿Cómo usan los geólogos las deducciones y las observaciones? (Cap. 24)
4. ¿Qué pruebas hay para sostener la teoría del desplazamiento continental (Cap. 24)
5. ¿Cuáles son los rasgos principales del interior de la Tierra? (Cap. 24)
6. Usa tu conocimiento de la estructura de la Tierra para explicar por qué el fondo del Océano Atlántico se está haciendo más ancho. (Cap. 24)
7. ¿Cuáles son algunos de los agentes de desgaste? (Cap. 25)
8. ¿Qué rol tiene el agua en el proceso de desgaste? (Cap. 25)
9. ¿Qué rol tiene la gravedad? (Cap. 25)
10. ¿Cuáles son algunas de las pruebas de la existencia de extensos glaciares en el pasado? (Cap. 26)

## Vocabulario

| | |
|---|---|
| **Asthenosphere** (463) | **Astenosfera** una capa de la corteza terrestre ubicada en el manto superior, directamente debajo de la litosfera; a causa de las altas temperaturas y la presión, esta área dura y rocosa puede fluir |
| **Continental drift** (460) | **Deriva continental** teoría de que los continentes se formaron cuando Pangaea, una gran masa de tierra, se dividió y sus pedazos comenzaron a alejarse unos de otros |
| **Core** (463) | **Núcleo** la porción central de la Tierra que está bajo el manto, comienza a una profundidad de más o menos 2900 km, tiene temperaturas que van de 2200 a 5000 grados centígrados, y probablemente consiste de hierro y níquel fundidos. Tiene un núcleo interior sólido y un núcleo exterior líquido. |
| **Crust** (463) | **Corteza** la capa exterior de la Tierra, fina y rocosa; también se conoce como la superficie terrestre |
| **Delta** (506) | **Delta** un área de sedimentos triangular o en forma de abanico depositada en la boca de un río |
| **Erosion** (487, S190) | **Erosión** el transporte de tierra, roca, y otros materiales en la superficie de la Tierra |
| **Estuary** (507) | **Estuario** una boca de río ancha y llana que en general se extiende mucho hacia dentro de la tierra y donde el agua fresca de un río se mezcla con el agua salada de un océano |
| **Geology** (453) | **Geología** el estudio científico del origen, la historia, y la estructura de la Tierra y de los cambios traídos por los procesos naturales |
| **Glacier** (512) | **Glaciar** una masa enorme de hielo que lentamente fluye sobre una gran área de tierra |
| **Gravity** (487) | **Gravedad** la fuerza de atracción entre los objetos |
| **Ground water** (497) | **Agua subterránea** agua que se junta bajo la superficie de la Tierra |
| **Gully** (504) | **Barranco** una cuneta profunda que ha cavado en la Tierra el agua que corre |
| **Hot spot** (468) | **Calor concentrado** un área donde se concentra el calor en lo profundo de la Tierra que causa que algunas de las rocas en el manto de la Tierra se fundan, formando una columna de magma que se eleva hacia la superficie |
| **Lithosphere** (463) | **Litosfera** la capa exterior de la Tierra, sólida y rocosa, que consiste de la corteza y del manto superior y que se divide en placas tectónicas |
| **Mantle** (463) | **Manto** la capa media de la Tierra, entre la corteza y el núcleo |
| **Meltwater** (517) | **Agua derretida** agua de nieve o hielo derretida; usualmente asociada con un glaciar que se está fundiendo |

**Mid-ocean ridge** (464) — **Risco oceánico** una de varias cadenas de montañas bajo el agua, formadas en conexión con centros de separación

**Outwash** (517) — **Arrastre** arena y grava que se han sacado de un glaciar y han sido depositadas más allá de los límites del glaciar por corrientes de agua derretida

**Pangaea** (460) — **Pangaea** de acuerdo a la teoría de Alfred Wegener, la masa de tierra única de la que salieron todos los continentes

**Permeability** (496) — **Permeabilidad** medida de cómo el agua, el aire, o las raíces de las plantas penetran o atraviesan un material poroso como la tierra

**Plate tectonics** (461) — **Tectónica de placas** la teoría de que la Tierra está dividida en un pequeño número de grandes placas rígidas que se mueven, causando la aparición de volcanes y terremotos y que resultan en cambios en la forma y el tamaño de las cuencas oceánicas, las montañas y los continentes

**Porosity** (496) — **Porosidad** medida del espacio vacío en un material, que es determinada por el volumen de agua que puede ser absorbido por el material

**Runoff** (495, S192) — **Escurrimiento** exceso de agua de lluvia que la tierra no absorbe, pero que comienza a juntarse y corre sobre la superficie de la tierra

**Sediment** (474) — **Sedimento** fragmentos de roca desgastada, como arena, barro, y pequeños pedazos de grava, que son trasladados y depositados por el viento, el agua, o el hielo

**Simulate** (484) — **Simular** representar, modelar, o imitar una situación, un sistema o un proceso

**Spreading center** (464) — **Centro de separación** una zona linear, usualmente encontrada en el medio del océano, donde dos placas tectónicas que se tocan se separan, y se abre un área a la cual fluye continuamente roca fundida del interior de la Tierra, creando una nueva corteza

**Tectonic plate** (463, S182) — **Placa tectónica** una de varias secciones grandes de la corteza terrestre. Todas estas secciones están en movimiento constante, de acuerdo a la teoría de tectónica de placas

**Terminal moraine** (517) — **Morena terminal** un risco de roca formado por los materiales que quedaron de un glaciar que ha dejado de avanzar

**Till** (515) — **Morena** un tipo de sedimento que quedó de los glaciares

**Trench** (464) — **Cauce** un valle largo, profundo, creado cuando chocan dos placas y la fuerza de la colisión empuja a una de las placas hacia abajo

**Tributary** (504) — **Tributario** arroyo o río que fluye hacia un río, lago, u otra corriente de agua más grande

**Watershed** (504, S192) — **Vertiente** región drenada por un río, sistema de ríos, u otra corriente de agua

**Weathering** (481, S185) — **Desgaste** un proceso químico o mecánico en el que las rocas expuestas a la intemperie son corroídas por el agua, el viento, o el hielo

## Photo/Art Credits

Abbreviated as follows: (t) top; (b) bottom; (l) left; (r) right; (c) center.

**Photo Credits**
Front Cover: (tl), Sinclair Stammen/Science Photo Library/Photo Researchers, Inc.; (br), Antony Mercieca Photo/Photo Researchers, Inc.; (bkgd), Page Overtures. Back Cover: (tl), Tony Stone Images; (br), Jeff Smith/FotoSmith/Reptile Solutions of Tucson; (bkgd), Page Overtures. Title Page: i(bkgd), Page Overtures; (bl), Jeff Smith/FotoSmith/Reptile Solutions of Tucson. Page 23, Ward's Natural Science Establishment; 28, Phil Degginger; 50, Tom Bean.

**Art Credits**
*All work, unless otherwise noted, contributed by Holt, Rinehart and Winston*
Page 6, Dodson Publication Services; 8, Morgan Cain & Associates; 10, Uhl Studio; 13 Dodson Publication Services; 16, Craig Attebury/Jeff Lavaty Artist Agent; 18, (t) Dennis Jones,(b) Valerie Marsella; 21, Dodson Publication Services; 34, David Merrell/Suzanne Craig Represents; 39, Morgan Cain & Associates; 49, Morgan Cain & Associates; 50, Uhl Studio; 51, Morgan Cain & Associates; 53, Maryland Cartographics; 55, (b) Richard Pembroke/American Artists; 57, Gary Locke/Suzanne Craig Represents; 59, Dodson Publication Services; 64, Lyne Raff; 66, Walter Stuart/Richard Salzman Artist Representative; 68, Morgan Cain & Associates; 71, Morgan Cain & Associates; 94,(t) Dodson Publication Services, (b) Uhl Studio; 96, Craig Attebury/Jeff Lavaty Artist Agent; 97, (t) Dennis Jones,(b) Valerie Marsella; 98, Dodson Publication Services; 102, David Merrell/Suzanne Craig Represents; 108,(t) Morgan Cain & Associates,(b) Uhl Studio; 110, Richard Pembroke/American Artists; 111, Gary Locke/Suzanne Craig Represents; 112, Dodson Publication Services; 114, Lyne Raff; 115, Walter Stuart/Richard Salzman Artist Representative; 116, Morgan Cain & Associates; 118, Morgan Cain & Associates.

## Answer Keys

# Unit 8: Our Changing Earth

Contents







Name ______________ Date ________ Class ________

Chapter 24
Activity Worksheet



## Earth Trivia

Complete this activity as you read Lesson 1, Seeing the Sights, which begins on page 453 of your textbook.

Check your knowledge of Earth facts by answering the questions below. You may find certain resources such as encyclopedias or almanacs helpful in your search.

**1.** The longest river in the world is 6670 km long. It is the
**a.** Mackenzie in Canada. **b.** (Nile in Africa.) **c.** Mississippi in the United States. **d.** Chang in Asia. **e.** Amazon in South America.

**2.** Mount Everest is the highest mountain peak in the world. Its height above sea level is approximately
**a.** 3 km. **b.** 5 km. **c.** 7 km. **d.** (9 km.) **e.** 11 km.

**3.** The deepest place on Earth is the Marianas Trench in the Pacific Ocean. Its depth below sea level is
**a.** 3 km. **b.** 5 km. **c.** 7 km. **d.** 9 km. **e.** (11 km.)

**4.** As of 1992, the tallest free-standing structure in the world was the CN Tower in Toronto, Canada. It is 550 m (more than 0.5 km) high. The highest natural waterfall in the world is Angel Falls in Venezuela. It is ________ the height of the CN Tower.
**a.** one-third **b.** one-half **c.** equal to **d.** (two times) **e.** three times

**5.** The length of the Rocky Mountains, which is the second longest range in the world, is
**a.** 2000 km. **b.** 4000 km. **c.** (6000 km.) **d.** 8000 km. **e.** 10,000 km.

**6.** The search for coal, natural gas, and oil leads us to dig deep into the Earth. How deep is the deepest mine, Western Deep in South Africa?
**a.** 0.5 km **b.** 1 km **c.** 2 km **d.** (3 km) **e.** 4 km

**7.** North America is a large continent. But how large is it? Among all continents in the world, North America ranks ________ in area.
**a.** first **b.** second **c.** (third) **d.** fourth **e.** fifth

**8.** Our planet is called "Earth," not "Water." Should it be? The surface is covered by both land and water. Which is the correct proportion of the two?
**a.** The areas of land and water are about equal.
**b.** (The area of land is one-half the area of water.)
**c.** The area of land is twice the area of water.
**d.** The area of land is three times the area of water.

**9.** The largest freshwater lake in the world (by volume) is in
**a.** the United States. **b.** Canada. **c.** (Russia.) **d.** Tanzania. **e.** Malawi.

**10.** The largest island in the world is
**a.** (Greenland.) **b.** New Guinea. **c.** Great Britain. **d.** Victoria Island (Canada). **e.** Ellesmere Island (Canada).

Name ______ Date ______ Class ______

## EXPLORATION 1

Chapter 24 Exploration Worksheet

### Selling Scenery, page 454

**Cooperative Learning Activity**

| | |
|---|---|
| **Group size** | 3 to 4 students |
| **Group goal** | to write a sales presentation or design a travel brochure for geologists |
| **Individual responsibility** | Each member of your group should choose a role such as artistic director, copywriter, team manager, or editor. |
| **Individual accountability** | Each group member should write a self-assessment to review his or her personal participation in the group and the effectiveness of his or her contributions. |

How good a salesperson are you? Test your skill. Choose *one* of the activities below.

1. You are a travel agent. What would you say to your clients to persuade them to visit the scenic landscape you chose in step 4 on page 453 of your textbook? Write down the main points you would make.
2. Design a travel brochure to persuade vacationers to visit the area you chose. Highlight the area's scenic attractions.
3. You are a travel agent again, but this time your client is a geologist. How would you change your sales pitch?
4. Now you are a geologist, and you are planning an expedition to a place of your choice. Where would you go? What would be your sales pitch to get others to join you?

Pitch your strategy in the space below. Feel free to use a separate piece of paper or any classroom materials to enhance your plans.

**Whether students choose to write a sales presentation or design a travel brochure, stress that their emphasis should be on geological features. Common tourist attractions such as restaurants, galleries, and museums may also be included. In question 3, students should focus strictly on geological attractions. Since this sales presentation is directed at a geologist, encourage students to use precise, scientific language.**







Name ______ Date ______ Class ______

## EXPLORATION 3

Chapter 24 Exploration Worksheet



### Creating a Model, page 462

| **Your goal** | to create a model to study plate movement by convection current | **Safety Alert!** |
|---|---|---|

Although questions still remain, most scientists think that plate movement is caused by convection currents. When a gas or a liquid is heated unevenly, the part that is heated rises. The movement of heated gas or liquid is called a *convection current.* To study processes that they cannot see, scientists often make models. In this Exploration, you will create a model of plate movement by convection current.

#### You Will Need

- water
- a shallow, rectangular pan (30 cm × 40 cm)
- a hot plate
- dark food coloring
- 8 pieces of thin cardboard, about 1 cm × 1 cm
- a watch or clock with a second hand

#### What to Do

1. Fill the pan three-quarters full of water. Place the pan on the hot plate.
2. Heat the water over very low heat for 30 seconds. Add a few drops of food coloring to the center of the pan. Write down your observations in the space provided. Turn off the heat.
3. Label the cardboard pieces 1 through 8. Carefully place pieces 1 through 4 in the center of the pan, as close together as possible. Place the rest of the cardboard pieces in the corners of the pan.
4. Turn the hot plate on low. In your ScienceLog, sketch the pattern of movement for each cardboard piece. Turn off the heat.

#### Thinking Things Over

1. What happened when you added food coloring to the water? How can you explain your observations?

   **The food coloring moved in a circular pattern. This was because of convection currents caused by heating the water in the pan.**

Name ______________ Date ________ Class ________

Exploration 3 Worksheet, continued

2. Describe what happened to the cardboard pieces when the water was heated.

**When the water was heated, the cardboard pieces in the center of the pan moved away from each other in a circular pattern. The pieces in the corners of the pan did not move at first unless they were touched by one of the other pieces. Eventually, all of the pieces began to move in circles.**

Do you think this activity is a good model for plate tectonics? Explain.

**Answers will vary. Students may note that in this model, the heat source was closer to some of the pieces than to others. Heat from the Earth's core is more evenly distributed than is the heat in the model.**

3. Examine the illustration of the Earth's interior below. What portion of the Earth does each of the following represent: the water, the cardboard pieces, the hot plate?

**The water represents the molten rock of the Earth's mantle, the cardboard pieces represent tectonic plates, and the hot plate represents the Earth's core.**



Illustration also on page 463 of your textbook







Name ______________ Date ________ Class ________

Chapter 24
Theme Worksheet



## Using the Theme of Energy

This worksheet is an extension of the theme strategy outlined on page 464 of the Annotated Teacher's Edition. It is also an extension of Continental Facts on page 464 of the Pupil's Edition.

| Focus question | How is the occurrence of an earthquake along a sliding plate boundary similar to the breaking of a stretched rubber band? | Safety Alert! |
|---|---|---|

### You Will Need

- two wooden blocks, 5 cm × 10 cm × 20 cm
- a rubber band, size 33

### What to Do

1. Lay the two wooden blocks side by side on a desk or the floor.
2. Position the rubber band around both wooden blocks as shown below.



3. Move one of the blocks slowly in one direction while moving the other block in the opposite direction. Describe what happens when the blocks are moved.

**The blocks move easily at first. As the blocks move further, however, the tension from the stretched rubber band makes the movement more difficult.**

Name ______________ Date ________ Class ________

Theme Worksheet, continued

4. Continue moving the blocks in opposite directions, slowly and steadily. Describe what happens.

**As the blocks move in opposite directions, the tension in the rubber band increases. It becomes very difficult to move the blocks further once the rubber band achieves its maximum tension. Eventually the stress is too great, and the rubber band breaks. At that point the blocks move easily past one another.**

5. As the rubber band stretches, energy is stored in the rubber band. When is this stored energy released?

**When the rubber band breaks**

6. Explain how this activity demonstrates the buildup and release of energy that occurs during an earthquake at a sliding plate boundary.

**Sample answer: Two sliding crustal plates (represented by the blocks) may not move smoothly past one another because of friction (represented by the tension of the rubber band). As the plates continue to try to move past one another, energy (represented by the increasing tension of the stretching rubber band) builds up in the plates. This energy is released as the plates along the plate boundary give way (the rubber band breaks) and the plates (the two blocks) suddenly move past one another to a new position. This release of energy takes the form of an earthquake.**





Name ______________ Date ________ Class ________



## Geo-Cross

Complete this activity before beginning Challenge Your Thinking on page 468 of your textbook.

Use the clues below to complete the crossword puzzle.

**Across**

1. A small aquatic lizard whose fossils are found in only two places in the world
2. Meaning "all Earth," this word is the name of a supercontinent that may have existed 300 million years ago
7. Plate interactions can cause these.
9. The outermost layer of the Earth; it is thin and rocky.
10. The Earth's center is known as this.
11. 1 down and 9 across together are known as this.
12. The layer of the mantle directly below the lithosphere
15. A scientist who studies the Earth
16. The only major plate that does not have a continent on it

Across answers: 1 MESOSAURUS, 2 PANGAEA, 7 VOLCANOES, 9 CRUST, 10 CORE, 11 LITHOSPHERE, 12 ASTHENOSPHERE, 15 GEOLOGIST, 16 PACIFIC

Down answers: 1 MANTLE, 2 PLATES, 3 MOUNTAINS, 4 EARTHQUAKES, 5 WEGENER, 6 TECTONICS, 8 LAVA, 13 TRENCH, 14 SEVEN

**Down**

1. The middle layer of the Earth, between the crust and the core
2. The Earth's surface is made up of a number of huge ________.
3. These are created when sections of the Earth's surface are folded and thrust upward.
4. These are caused by sections of the Earth's surface sliding past each other, often violently.
5. This man hypothesized the existence of 2 across.
6. The idea that the Earth's surface is made of moving pieces is called the theory of plate ________.
8. This is often expelled from the eruptions of 7 across.
13. A deep valley caused when one section of the Earth's surface is forced downward
14. There are ________ major plates on the Earth's surface.

Name ______________ Date ______ Class ______

**Chapter 24**
Review Worksheet

## Challenge Your Thinking, page 468

**1. Island Hopping**

A *hot spot* is an area of concentrated heat, deep within the Earth's interior. Hot spots melt the rocks of the mantle, which creates a column of magma that rises toward the surface. Examine the map of the Hawaiian Islands below.

**a.** Use the map and your knowledge of the theory of plate tectonics to explain how the Hawaiian Islands might have formed.

**The hot spot beneath the Earth's surface is causing magma to be forced to the surface. This hot spot remains in place while the plate is carried over it, resulting in a series of islands.**

**b.** Which islands are the oldest? How can you tell?

**The islands in the upper left of the diagram are the oldest because they are the farthest away from the hot spot. The movement of the plate has taken the islands that were formed first away from the hot spot.**

**c.** Do you think any new islands will form? Why?

**New islands will continue to form as long as the Pacific plate continues to move over the hot spot.**

**d.** Which islands are most likely to have active volcanoes? Explain.

**The youngest islands are most likely to be volcanically active because they are nearest to the hot spot.**

Kauai
Oahu
Molokai
Maui
Hawaii
Oceanic crust
Magma
Direction of plate movement
Hot spot

Illustration also on page 468 of your textbook





Name ______________ Date ______ Class ______





**2. A More Persuasive Argument**

Using the information contained in this chapter, revise the letter that you wrote to the Royal Geological Society in Exploration 2 on page 459 of your textbook. Carefully organize any additional evidence to make your argument as persuasive as possible. Continue in your ScienceLog if necessary.

*April 1, 1911*

*Royal Geological Society*
*Holywell House, Worship Street*
*London, EC2A2EN*

*My Esteemed Colleagues,*

*I wish to bring to your attention a matter of great importance. For several months I have been working on an idea . . .*

**In their revised letter, students could include some of the following information in addition to the evidence given in Exploration 2 on page 459:**

- **Identical rock formations and fossils are found in Africa and South America.**
- **Coal deposits in Antarctica indicate that this continent once had a warmer climate and may have once been closer to the equator than it is now.**
- **The fossil plant *Glossopteris* is found in southern South America, southern Africa, India, Australia, and Antarctica.**
- **The Atlantic Ocean between North America and Europe widens by 1 cm each year. The creation of new crust along the sea floor is marked by underwater mountain chains called mid-ocean ridges.**
- **The boundaries between tectonic plates are often marked by deep trenches, mountains, and volcanoes.**

Letter also on page 459 of your textbook

Name ____________ Date ________ Class ________

Chapter 24 Review Worksheet, continued

**3. Lizard Debate**

Fossils of the reptile *Mesosaurus* are found only in South America and Africa. Wegener used this evidence to support his theory of continental drift. Imagine that you are a scientist who refuses to accept Wegener's theory. Suggest other explanations for the location of *Mesosaurus* fossils. They certainly didn't have speedboats! Can you come up with a more reasonable explanation?

**One possible explanation for the location of *Mesosaurus* fossils is that a land bridge connected the continents of Africa and South America at one time.**

Illustration also on page 468 of your textbook

**4. Photo Mystery!**

Carefully examine the photo at the top of page 469 of your textbook. Make inferences about what caused the scene. How would you check the accuracy of your inferences?

**Sample answer: The curved layers of rock indicate that the crust in this location has been folded by the collision of continental plates. A geologist could be consulted to determine the accuracy of this inference.**

**5. Sunken Treasure**

Imagine that you are a travel agent. You are offering submarine tours of a spreading center at the bottom of the ocean. In your ScienceLog, write a brochure or sales pitch designed to convince an average tourist (not a geologist) to take this tour.

**Answers will vary but should demonstrate student understanding of spreading centers. Encourage students to be creative in the delivery of their sales pitch.**

Illustration also on page 469 of your textbook





Name ____________ Date ________ Class ________

Chapter 24 Assessment



## Correction/Completion

1. The following sentences are incorrect or incomplete. Make the sentences correct and complete.

 a. The mantle and the crust make up the portion of the Earth known as the asthenosphere.

 **The mantle and the crust make up the portion of the Earth known as the *lithosphere*.**

 b. The lithosphere is divided into segments known as **tectonic** plates.

## Short Responses

2. Place a check mark beside each observation that supports the theory of continental drift.

 a. ✓ The continents look somewhat like puzzle pieces that fit together.

 b. ____ Portuguese is spoken by most people in Brazil and in Portugal.

 c. ✓ Rock formations found in Africa are identical to formations found in South America.

 d. ____ Scientists have discovered dinosaur fossils that are 200 million years old.

3. Match each type of plate activity with a geological event.

 a. separating — **c** Earthquakes occur because of friction between plates.

 b. converging — **a** New crust is formed as molten rock flows into the gap between two plates.

 c. sliding — **b** Trenches are formed in the ocean floor.

Name ______________ Date ________ Class ________



## Short Essay

**4.** Explain how deep trenches in the ocean floor and nearby mountain peaks are related to each other.

**Sample answer: Both are the result of colliding tectonic plates. When an oceanic plate meets a continental plate, the oceanic plate is driven downward to form a trench, and the continental plate is folded and lifted to form new mountains.**

## CHALLENGE 1 Word Usage

**5.** One definition of *tectonic* is "having to do with building or construction." How does this definition relate to what you've learned about the theory of plate tectonics?

**Sample answer: Tectonic plates can be thought of as the building blocks of the Earth's crust. According to the theory of plate tectonics, continental plates are separate pieces that move and change position on the surface of the Earth. Their interactions with each other can create various physical features, constantly changing the appearance of the Earth's surface.**





Name ______________ Date ________ Class ________





## CHALLENGE 2 Illustration for Interpretation

1
Plate movement
Hot spot

1 2
Plate movement
Hot spot

1 2 3
Plate movement
Hot spot

**6. a.** Examine the series of diagrams above and explain what process is described by them.

**Sample answer: A volcanic island forms over a hot spot. As the plate moves, it carries the island with it, and another island is formed over the hot spot. The plate continues to move, and another island is formed.**

**b.** Which island is the oldest? **1**

**c.** Do the diagrams support or disprove the theory of plate tectonics? Explain.

**The diagrams support the theory because they explain how islands are formed by plate movement.**

Name ______ Date ______ Class ______

Chapter 25
Discrepant Event Worksheet

## Penny Protection

Conduct the following experiment as you begin Chapter 25, which starts on page 470 of your textbook.

### You Will Need

- a large aluminum pie plate
- a hose with a sprayer attachment
- 5 or 6 pennies
- a small mound of pottery clay (do not use modeling clay or other oily clays)

### What to Do

1. Form a flat brick out of the clay, about 15 cm × 5 cm × 3 cm. Place the clay in the aluminum pie plate. Then firmly press the pennies into the top of the clay. Make sure that the pennies are spaced evenly along the top of the clay brick.
2. Take the pie plate outside.
3. What do you predict would happen if you directed a stream of water from the hose at the pennies?
   **Expected answer: The force of the water from the hose would knock the pennies out of the clay.**
4. Now spray the brick with water from the hose. Begin by using a gentle spray and then increase the pressure until the brick begins to wash away.

### Questions

1. What happens to the clay when the hose is turned on?
   **The water washes away some of the clay.**
2. What do you observe about the clay underneath the pennies?
   **The clay does not wash away or does not wash away as quickly.**
3. Why do you think this happened?
   **The pennies protected the clay underneath them from the water.**
4. Imagine that the pennies are actually a very hard rock, such as granite, and that the clay is a very soft rock such as limestone. In your ScienceLog, draw a diagram showing what the rock formation might look like before and after the effects of the water.
   **Diagrams should illustrate the principle demonstrated by the clay and the pennies. For example, the granite would protect the limestone from damage caused by water just as the pennies protected the clay. The limestone not covered by granite is at a higher risk of being washed away, just as the clay not covered by the pennies was quickly washed away by the water.**





Name ______ Date ______ Class ______

Chapter 25
Exploration Worksheet

## EXPLORATION 4

## Monuments to Change, page 483

**Your goal** to learn about the relationship between weathering and time

If there is a graveyard in your community that is over 100 years old, you have the makings of an interesting project.

### Problem 1

What is the relationship between the amount of weathering and the length of time that a monument has been exposed? Hint: Find stones of different ages made from the same kind of rock. Look for newer ones (0–100 years old), older ones (over 150 years old), and some in between. Compare the amount of weathering in each. Keep track of your findings in your ScienceLog, and then summarize your findings here.

**Generally, the longer a gravestone has been exposed, the greater the amount of weathering that has occurred.**

### Problem 2

Do different types of stones weather at different rates? Hint: Find several gravestones of approximately the same age but made of different kinds of rock. Compare the amount of weathering in each.

**Harder rock, such as granite, will probably show less weathering than rock made of marble, even if the ages of the rocks are the same.**

Photo also on page 483 of your textbook

What do you conclude from your study?

**Students may conclude from Exploration 4 that gravestones exposed for longer periods of time will show more weathering. However, hard rocks, such as granite, weather more slowly than soft rocks, such as marble.**

Name ______________ Date ________ Class ________

# EXPLORATION 5

Chapter 25
Exploration Worksheet

## Three Simulations, page 484

### Cooperative Learning Activity

| | |
|---|---|
| **Group size** | 3 to 4 students |
| **Group goal** | to simulate factors that might affect the speed of weathering |
| **Individual responsibility** | Each member of your group should choose a role such as leader, materials manager, safety monitor, or recorder. |
| **Individual accountability** | Each group member should be able to express what he or she learned by performing the simulations. |

It is very difficult (not to mention very time consuming) to determine the speed of weathering through actual observation. Real rocks take a very long time to break down. Sometimes, however, laboratory activities can *simulate* (imitate) what happens in the environment. Three factors that might affect the speed of weathering are investigated in the simulations that follow.

### You Will Need

- a magnifying glass
- rock samples (granite, limestone, sandstone, brick, and a piece of concrete)
- a tray
- a freezer

### Simulation 1

What effect do freezing and thawing have on rocks? Make a prediction.

**Predictions will vary; accept all reasonable responses.**

### What to Do

1. Examine your samples with the magnifying glass. Can you find any cracks where water might enter?
2. Soak each rock sample in water. After several minutes, remove it and place it on a tray in the freezer overnight.
3. The next day, remove the samples from the freezer, and allow the rocks to warm up to room temperature.
4. Examine the samples for any changes. Then repeat the process of soaking, freezing, and thawing. If time allows, repeat the freezing-thawing cycle several times.

### Questions

1. Compare what happens to the different samples. Which sample shows the greatest change? the least change?

   **Sandstone and concrete should show the greatest change, while granite should show the least.**

Chapter 25





Name ______________ Date ________ Class ________

Exploration 5 Worksheet, continued

2. How does this simulation relate to conditions in the environment?

   **The activity simulates the repeated processes of freezing and thawing caused by yearly weather cycles.**

### You Will Need

- a whole piece of chalkboard chalk
- dilute hydrochloric acid
- a graduated cylinder
- two 250 mL beakers
- latex gloves

### Simulation 2

### What to Do

1. In addition to your safety goggles, wear latex gloves and an apron during this activity. Break a piece of chalkboard chalk in half. Put aside one half, and break the other half into several small pieces. (You can wrap it in a paper towel and hit it gently with a hammer.)
2. Carefully pour 100 mL of dilute hydrochloric acid into each beaker.
3. Place the large piece of chalk in one beaker and the smaller pieces in the other.
4. Clean up your lab area and wash your hands when you are finished.

### Questions

1. In which beaker did the bubbling last the longest?

   **The bubbling lasts the longest in the beaker with the large piece of chalk.**

2. What happened to the chalk in each beaker?

   **In both beakers, some chalk may disappear. More chalk disappears in the beaker with smaller pieces.**

Name ____________ Date ________ Class ________

Exploration 5 Worksheet, continued

3. What comparison can you make between this investigation and the natural weathering process?

**In both this investigation and the natural weathering process, the rate of chemical weathering is faster for smaller rocks than for larger rocks of the same material. This is because the rate depends on how much surface area is exposed.**

## Simulation 3

Do some kinds of rocks resist weathering better than others? One class investigated this problem by using marble chips and quartzite. They took out 100 g of each rock and placed each in a container about half full of water. After putting on the cover, they shook each container for 5 minutes at a rate determined before the experiment. After draining each container, they weighed and recorded the mass of the chips and quartzite. They repeated the procedure three more times and found the mass of each sample after each 5-minute period of shaking, for a total of 20 minutes. Their results are recorded in the table below.

| Time (min.) | Mass of marble chips remaining (g) | Mass of quartzite chips remaining (g) |
|---|---|---|
| 0 | 100.0 | 100.0 |
| 5 | 98.5 | 99.7 |
| 10 | 96.9 | 99.6 |
| 15 | 95.3 | 99.3 |
| 20 | 92.7 | 99.2 |

### Questions

1. Which kind of rock lost the greatest amount of mass?

**Marble lost the most mass.**

2. What inferences can you make about the resistance of marble to weathering? the resistance of quartzite to weathering?

**Quartzite is more resistant to weathering than marble is.**







Name ____________ Date ________ Class ________

Exploration 5 Worksheet, continued

3. During which time period did the greatest loss of mass from the marble chips occur? Can you explain this?

**The greatest loss of mass occurred between 15 and 20 minutes because more surface area was exposed as the rock broke into smaller pieces.**

4. What do you think might happen if you shook each sample for 1 hour?

**The mass of the remaining marble chips in each sample would continue to decrease.**

5. How does this investigation simulate the weathering process?

**It shows how rocks can be eroded by moving water and by collisions with one another.**

### How Fast Does Weathering Occur in Your Community?

Weathering occurs at different rates, depending on the conditions in the environment. Suggest conditions in the environment that affect the speed of weathering (other than those you have already dealt with, of course). Make a list of all the factors in your area that might affect the speed of weathering.

**Answers will vary according to local conditions but should include a combination of physical and chemical factors. Student lists might include the following factors: the types of materials found in their community; the frequency of frost; the level of humidity, temperature, and wind and water movement; air pollution and other factors related to human activity; and physical and chemical factors related to the activity of plants and animals.**

Name ______________ Date ________ Class ________

# EXPLORATION 6

Chapter 25
Exploration Worksheet

## A Model Landslide, page 489

**Cooperative Learning Activity**

| | |
|---|---|
| **Group size** | 3 to 4 students |
| **Group goal** | to investigate a different variable of the landslide model and to demonstrate the result to the rest of the class |
| **Individual responsibility** | Each member of your group should choose a role such as facilitator, materials coordinator, reporter, or timekeeper. |
| **Individual accountability** | Each group member should be able to list the variable that his or her group investigated and give a brief summary of the group's results and conclusions. |

Each year one of the projects in Mrs. Jefferson's class is to build a model and do an experiment with it. Pam and Sanjay decided to build a model that would help them figure out what causes landslides. Pam and Sanjay thought of the following ways to get the sand to slide:

1. Pour water on the sand.
2. Pile the sand deeper.
3. Make the angle of the board steeper.



### You Will Need

- a shallow, rectangular pan
- sand
- water
- a piece of plywood
- wooden blocks

### What to Do

1. Test Pam and Sanjay's methods. (You can see their model on page 489 of your textbook.) What other methods might they have tried?
   **Sample answer: jarring the board, pushing on the sand, or even using material other than sand (clay, topsoil, or a mixture of sand and gravel)**
   Do all these methods cause the sand to slide down the board?
   **Yes; all of these methods should cause the sand to slide down the board.**
2. Can you think of any natural occurrences that match your investigations?
   **Sample answers: Pouring water on sand matches soil saturation after a rain; piling on more sand matches the buildup of sand, snow, or loose rocks; and changing the angle of the board matches the natural differences that exist in the degrees of slopes. More possible answers are listed on page 489 of the Annotated Teacher's Edition.**





Name ______________ Date ________ Class ________

Chapter 25
Review Worksheet

## Challenge Your Thinking, page 491

**1. Artist's Choice**

You are a sculptor, and you have been asked to design a monument that will be placed in downtown New York City. How would you go about picking the type of stone that would be best to use? What would you have to know?

**A sculptor would need to research the type of climate there, including humidity, temperature changes, and wind, as well as the amount and type of pollution. The sculptor would also need to find out which types of stone resist weathering the best.**

Illustration also on page 491 of your textbook

**2. Physical or Chemical?**

There are two types of weathering: physical and chemical. Limestone dissolving to form caves is an example of chemical weathering. The cracking of rock during a freeze-thaw cycle is an example of physical weathering. Classify the following examples of weathering as *chemical, physical,* or *not enough information to say.*

| | |
|---|---|
| **Physical** | A rock splits open after a day in the hot sun. |
| **Not enough information to say** | A block of sandstone grows soft and crumbly after being exposed to water for a period of time. |
| **Physical** | A rock breaks into pieces when it is heated. |
| **Chemical** | A rock develops a grayish color after being left outside. |
| **Physical** | A rock is ground down during a flash flood. |
| **Chemical** | A rock's surface is crumbly where a lichen grew on it. |

Name ____________ Date ________ Class ________

**Chapter 25 Review Worksheet, continued**

**3. Please Publish**

In your ScienceLog, write a human-interest magazine article entitled "I Survived the Frank Landslide" or "Buried Alive in an Avalanche!" Be sure to include a description of the cause of the disaster.

**4. Time Warp**

After a 20-year absence, Rip returns to his hometown. He is dumbfounded. "Why, it's just as I left it. It hasn't changed a bit. The old sycamore tree we used to have the swing on is still there by the old stone church. Look—the tree still has my initials scratched in it. It hasn't changed and neither has the church. The creek is still crystal clear. It hasn't changed . . . " Were things truly just as Rip left them, or should he take a closer look? What would Rip see if he looked more closely?

**Rip could have noticed whether the swing had been damaged, whether his initials were more difficult to read, whether the stone church showed signs of erosion, how much the tree had grown, and whether the creek had changed in shape or size.**

**5. Designer Display**

Design a bulletin-board display whose theme is geological changes. Ask yourself the following questions:

- Who is my audience?
- What information do I want to convey?
- How can I best convey the information—through articles? through photographs? through drawings?
- How can I make the display interactive enough to keep my audience interested?

**Encourage students to focus on conveying information effectively. Students may wish to use the Explorations in this chapter to generate ideas for their displays.**







Name ____________ Date ________ Class ________

**Chapter 25**
Assessment

**Word Usage**

1. In addition to the fact that they all start with *w*, what do the words *wind, wear, water,* and *weathering* have in common?

   **Sample answer: All four words can be used to describe the processes of erosion. *Weathering* occurs when forces such as *water* and *wind wear* away rocks and other features on the Earth's surface.**

**Correction/ Completion**

2. The following sentences are incorrect or incomplete. Make the sentences correct and complete.

   a. Erosion can be caused by rain, wind, glaciers, and other agents, but all forms of erosion involve **gravity**.

   b. Trees and other plants can prevent weathering by breaking up rocks, and they can cause erosion by holding soil in place.

   **Trees and other plants can *cause* weathering by breaking up rocks, and they can *prevent* erosion by holding soil in place.**

**Numerical Problem**

3. Every 2000 years, about 6 cm of land is removed from the continents by erosion.

   a. How much is removed in 1 million years? Show your work.

   **1,000,000 years/2000 years = 500; 500 × 6 cm = 3000 cm of land**

   b. How much is removed in 3 million years? Show your work.

   **3000 cm × 3 = 9000 cm of land**

   c. Why are geologists interested in this kind of data?

   **Sample answer: The amount of land removed by erosion over time is of interest to geologists because information like this can give insight into the age, history, and long-term characteristics of land structures.**

Name ________________ Date ________ Class ________

Chapter 25 Assessment, continued

CHALLENGE 1

### Short Essay

4. Could certain human activities affect the rate of weathering in an area? Explain.

**Students should recognize that human activities do impact the rate of weathering in the environment. There are many possible examples of this relationship. For example, certain farming practices can contribute to the erosion of topsoil. Plowing produces a loose surface layer of soil that is easily blown away by wind or washed away by rain. Clearing forests to produce lumber is another example of human activities that can contribute to soil erosion. Trees and shrubs absorb large quantities of water. When they are cut down, the amount of water running off the surface of the soil increases, and more soil is carried away. Sometimes the soil washes away slowly, and sometimes large amounts wash away in landslides. Pollution from automobiles and industry can also create chemicals that weather statues and buildings and kill vegetation.**

CHALLENGE 2

### Answering by Illustration

5. Suppose that you were given the supplies to build a medium-sized home of your own design. Design a structure that will resist the effects of weathering and erosion. Label the weather-resistant features.

**Answers will vary but should show a structure that includes features obviously designed to resist weathering. For example, it may show a sloped roof to prevent water from building up and causing damage. It may also include steel, acrylic plastic, bricks, or other materials that are known to be durable.**







Name ________________ Date ________ Class ________

Exploration 1 Worksheet, continued

### Making Sense of Your Results: Porosity and Permeability

Certain soils allow water to pass through them faster than others do. In other words, some soils are more *permeable* than others. When you measure the time it takes for water to pass through a soil, you are measuring the **permeability** of the soil. Which sample was the most permeable? the least permeable?

**Sand was the most permeable. Clay was the least permeable.**

Because of gravity, rainwater moves downward into tiny spaces in soil and rock. These spaces are called pores. The greater the volume of water that a soil can hold, the more *porous* the soil is. The volume of water that a soil can hold is a measure of the soil's **porosity.** Which sample was the most porous? the least porous?

**Garden soil was the most porous. Clay was the least porous.**

Which type of soil would cause the most runoff during a rainstorm? Explain your answer based on the concepts of porosity and permeability.

**Clay would cause the most runoff during a rainstorm. Since clay is the least permeable, it does not allow much water to soak into the ground. Since clay is also the least porous, it is unable to absorb much water. The excess rainwater becomes runoff.**

Name ______ Date ______ Class ______

EXPLORATION 2

Chapter 26
Exploration Worksheet

## Fast or Slow Streams, page 500

| Cooperative Learning Activity | | Safety Alert! |
|---|---|---|
| **Group size** | 3 to 4 students | |
| **Group goal** | to investigate the movement of water through different types of soil | |
| **Individual responsibility** | Each member of your group should choose a role such as timer, materials manager, results tracker, or experiment coordinator. | |
| **Individual accountability** | Each group member should be able to answer the following question: What determines whether a stream flows quickly or slowly? | |

### You Will Need

- a setup such as the one shown on page 500 of your textbook (Be sure that there is a drain at the lower end of the pan.)
- a metric ruler
- a wax pencil
- food coloring
- a watch with a second hand

### Activity 1

What effect does changing the slope of a stream have on the speed of the water?

### What to Do

1. Write down a prediction for the question above.

   **Accept all reasonable predictions.**

2. Set up the pan of water and the siphon as shown on page 500 of your textbook. Adjust the trough so that the upper end is 4 cm high.
3. With a wax pencil, mark a starting line on the side of the trough.
4. Open the siphon. Add a drop of food coloring to the water as it flows past the starting line.
5. Time how long it takes the dye to reach the end of the trough.
6. Change the height of the trough to 8 cm. Then repeat steps 4 and 5. In your ScienceLog, compare the results.







Name ______ Date ______ Class ______



### Activity 2

What effect does changing the volume of the water in a stream have on the speed of the water?

### What to Do

1. Write down a prediction for the question above.

   **Accept all reasonable predictions.**

2. Set up an experiment to answer this question. Consider the following as you plan what to do:
   - Can you use the same setup as before?
   - How can you increase the volume of water?
   - How will you determine the speed of the water?
   - Will you change the slope, as you did in Activity 1?
   - How will you determine whether the volume affects the speed of the water?
3. Can you think of another setup that might be used to answer this question?

   **Students may increase the volume of flowing water either by adding another siphon tube or by decreasing the size of the stream channel.**

### Analysis Please!

1. Look at the predictions you made. Does the data that you recorded support your predictions? In your ScienceLog, summarize your data, and present your conclusions.
2. You have identified two variables that affect how fast water flows in a trough. What other variables might affect the flow of a stream in a natural setting?

   **Sample answers: the shape of the stream bed, the amount and type of sediment carried by the stream, and the roughness of the stream bed**

3. Which stream do you predict can carry the most sediment, a fast-flowing stream or a slow-flowing one? What evidence do you have to support your prediction? Explain why the speed of a stream affects the amount of sediment that it can carry.

   **In general, a fast-flowing stream can carry more sediment than a slow-flowing one because the fast-flowing stream has more energy. Some students may have noticed that sediment accumulates in low-lying areas after a storm. Other evidence includes the formation of deltas at the mouths of rivers.**

Name ______ Date ______ Class ______

## Exploration 3

Chapter 26
Exploration Worksheet

### A Stream-Table Experiment, page 503

**Cooperative Learning Activity**

| | |
|---|---|
| **Group size** | 3 to 4 students |
| **Group goal** | to conduct an experiment and to gather data on stream formation |
| **Individual responsibility** | Each member of your group should choose a role such as primary investigator, supportive assistant, materials manager, or recorder. |
| **Individual accountability** | Each group member should be able to answer the questions posed on page 502 of his or her textbook and to answer the question asked at the end of this Exploration. |

**Safety Alert!**

### You Will Need

- a setup as shown in the photograph on page 503 of your textbook
- sand
- a metric ruler
- food coloring

### What to Do

1. Place sand to a depth of 5 cm in the stream table.
2. Wet the sand so that it can be shaped. Form the sand into a gentle slope.
3. Trace an S-shaped path with your finger almost to the bottom of the sand.
4. Turn on the water and adjust the flow so that it moves slowly into the tray.
5. Allow the water to flow down the path for 20 minutes. While the water is flowing, note the following locations:
   a. places where the water slows down or speeds up (food coloring may help here)
   b. places where banks are being eroded
   c. places where sand buildup occurs
   d. changes in the course of the stream
6. Turn off the water, but leave the sand in place.

Pretend that a friend was absent from school during the stream-table experiment. In your ScienceLog, describe to your friend exactly what you saw during the experiment. Do your observations support your choice of lot?
**Sample answer: We poured water down an S-shaped channel that we had dug down a sandy slope. The water slowed down around curves and sped up along straight areas. The water picked up sand from the outside of the curves and deposited sand on the inside of the curves. This caused the water channel to get wider and deeper and to curve even more. Students may conclude that lots C and D on page 502 of the textbook are also good choices.**





Name ______ Date ______ Class ______

## Exploration 4

Chapter 26
Exploration Worksheet

### Check It Out, page 505

**Your goal** to examine the effects of rivers

1. Is there a river near you? What stage is it in?
   **Answers will vary depending on your local geography.**

2. Many states contain well-known valleys or canyons. Find out about features such as these in your state. Then use a map to find out whether or not rivers flow through them. Try to answer these questions:
   a. Examine the photograph of the Grand Canyon on page 505 of your textbook. Which came first, the canyon (or valley) or the river?
   **The Colorado River came first. It began forming the Grand Canyon approximately 6 million years ago. Because the Colorado River flows through a hot, dry climate, the effects of weathering occur slowly.**
   b. What is the relationship between a valley, a canyon, and a river?
   **A young river erodes its channel, forming a shallow valley. As the river matures, the valley widens and deepens, forming a canyon.**
   c. A river may have flowed through a valley or canyon at one time, even though there isn't a river flowing there now. How can you tell?
   **The valley or canyon may follow a narrow channel where the river once flowed. There may also be accumulated sediments, remains of marine organisms, or markings on the valley or canyon walls indicating the effects of erosion.**

Name ______________ Date ________ Class ________

Chapter 26
Math Practice Worksheet

## The Grand Canyon Versus the Mighty River

Do this activity after reading Deltas and Estuaries, on page 506 of your textbook.

Imagine that you are a geologist and that you come across the following excerpt from a geological journal:

**EARTH ALERT:** A gradual change in the global climate is causing the Colorado River to slowly deposit sediment in the Grand Canyon. Scientists estimate that the present rate of deposition is raising the canyon floor by 0.05 mm per year.

Your geological interests lead you to ask some questions. Suppose that you've organized your questions and concerns into the following itemized list. Using your mathematical knowledge, answer the following five items about the fate of the Grand Canyon.

**Item 1** If the canyon is 1500 m deep, how long will it be until the river completely fills the canyon with sediment? Show your work.

**The canyon is 1500 m, or 1,500,000 mm, deep. If the rate of sediment deposition is 0.05 mm/yr., then it would take 1,500,000 mm ÷ 0.05 mm/yr., or 30,000,000 years, for the canyon to be completely filled with sediment.**

**Item 2** In the space below, graph the amount of sediment deposited in the Grand Canyon over an average human lifetime (about 75 years in the United States). Make sure to label the *x*- and *y*-axes and to give your graph a title.

**Sample answer:**

Sediment Deposition in the Grand Canyon

Sediment deposited (mm): 0.0, 0.5, 1.0, 1.5, 2.0, 2.5, 3.0, 3.5, 4.0

Time (years): 0, 5, 10, 15, 20, 25, 30, 35, 40, 45, 50, 55, 60, 65, 70, 75





HRW material copyrighted under notice appearing earlier in this work.

Name ______________ Date ________ Class ________

Math Practice Worksheet, continued

**Item 3** An animal is buried by the sediment at the bottom of the canyon. If a fossil hunter finds the animal 1 million years after the canyon is completely filled, how old is the fossil? Explain your answer.

**The animal is buried in the sediment at the bottom of the canyon, so (based on Item 1) its fossilized form would be 30 million years old by the time the canyon completely filled with sediment. If a fossil hunter finds the fossil 1 million years after the canyon is filled, the fossil would be 31 million years old.**

**Item 4** What will be the total volume of sediment dumped into the canyon if the canyon is 2 km wide and 30 km long? Show your work.

**The dimensions of the canyon are 2 km × 30 km × 1.5 km (the depth of the canyon is 1500 m, or 1.5 km, from Item 1). So the volume of the canyon, or the sediment that fills up the canyon, is 90 $km^3$.**

**Item 5** If the same amount of sediment from Item 4 were instead carried to the mouth of the river and dumped into the ocean, it would form a delta. Assuming that the ocean is 500 m deep at the mouth of the river and that the elevation of the new land at the delta is at sea level, what will be the surface area of the delta? (Hint: volume/depth = area) Show your work.

**Volume of sediment/depth of ocean = area of delta; 90 $km^3$/0.5 km = 180 $km^2$**

Name ______________ Date ________ Class ________

# EXPLORATION 5

Chapter 26
Exploration Worksheet

## How Do Glaciers Form? page 513

**Your goal** to learn how snow turns into ice

0 days

2 days

1 year

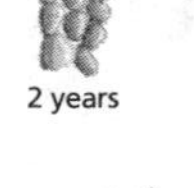

2 years

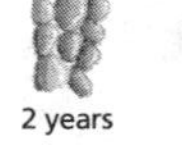

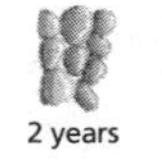

5 years

Illustration also on page 513 of your textbook

1. Can you turn snow into ice? How would you do it? It happens all the time in glaciers, although it takes quite a while.
2. To understand the process by which glaciers form, read the sentences below. Beware, though—their order is scrambled! Try to rearrange them into the proper order. Use the accompanying diagram as a guide.
   - **a.** The snow particles become rounded pellets due to the added weight of accumulating snow.
   - **b.** Evaporation and melting cause the flakes to lose their shape.
   - **c.** As the snow builds up, the pressure from the top layers squeezes out the air between the rounded pellets of snow. Icy particles called *névé* (NAY vay) are formed.
   - **d.** Star-shaped flakes collect in a fluffy mass with a lot of air caught between them.
   - **e.** The névé undergoes further packing and melting until it gradually changes into solid ice.

   **The correct order is as follows: d, b, a, c, e.**
3. Did you get the sentences in the right order? If you did, then you can truthfully say that you have got glaciers down cold!

Photo also on page 512 of your textbook





Name ______________ Date ________ Class ________

Chapter 26
Activity Worksheet

## Glaciers

Perform this activity after completing Exploration 5, How Do Glaciers Form? on page 513 of your textbook.

The illustrations below represent the changes in the size of a glacier over 50 years. Use the diagrams to answer the questions below.

After 10 years

After 20 years

After 30 years

After 40 years

After 50 years

1. What happened to the glacier from 10 to 20 years?

   **The glacier increased in size.**
2. What might have caused this change?

   **More snow accumulated during that time, and the weather may have been cooler.**
3. What happened to the glacier from 20 to 50 years?

   **The glacier decreased in size.**
4. What might have caused this change?

   **There was less snowfall during that time, and the weather may have been warmer.**

Name ______________________ Date __________ Class __________

Chapter 26
Exploration Worksheet

# EXPLORATION 6

## Can Ice Really Flow? page 514

**Your goal** to determine whether ice can flow

Can you design an experiment to prove that ice can change its shape or flow without melting? Perhaps you could try this idea.

1. Freeze a layer of ice about 1 cm thick in a tray. Once it has frozen, remove the ice from the tray.
2. In a freezer, support the layer of ice as shown on page 514 of your textbook. Hang a fairly heavy mass from the middle of the ice with a wire.
3. Observe what happens during the next few days.
4. Explain what you observe.

**Students should observe that the mass will slowly pass through the layer of ice without visibly affecting the shape of the ice.**

5. Repeat the experiment with different sized masses.
6. Can you apply your findings to explain how glaciers move?

**The force that gravity exerts on the mass determines the flow of the ice, just as the force that gravity exerts on the mass of a glacier determines the flow of the glacier.**





Name ______________________ Date __________ Class __________

Chapter 26
Activity Worksheet

## Water Words

Try this activity before moving on to Challenge Your Thinking, on page 520 of your textbook.

Find each of the words listed here in the word-search puzzle below. The words can be found by searching up, down, across, and diagonally. When you find a word in the puzzle, circle it.

| | | | |
|---|---|---|---|
| delta | iceberg | porosity | snow |
| esker | meltwater | rain | stream |
| estuary | mouth | river | *Titanic* |
| fjord | outwash | runoff | tributary |
| glacier | permeability | sediment | watershed |

```
B I H Q I Q V F G T X Y I O Y
N T P E R M E A B I L I T Y F
H N Q X B H O E P M X M E R Y
B E Q D Y Q V I X M A S R A G
X M R U N O F F D E T R E T R
I I I H N X C N R U P E K U E
R D V Q E N H T A L O I S B B
E E R A I N S R Y C R C E I E
T S M Q Y X Y D V I O A M R C
A X K O M M J W O N S L O T I
W Y P L H V M K R A I G U D U
T R O U T W A S H T T L T T K
L D E L T A S T N I Y H H D G
E N Z D E H S R E T A W K X L
M R I V E R F J O R D Q B U S
```





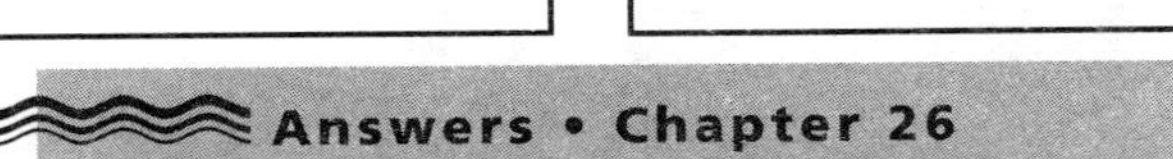

Name ______________ Date ________ Class ________

**Chapter 26**
**Review Worksheet**

## Challenge Your Thinking, page 520

### 1. Take a Hike!

If you followed a river upstream, you would notice that the river forks again and again. However, rivers almost never fork as they flow downstream. Why is this?

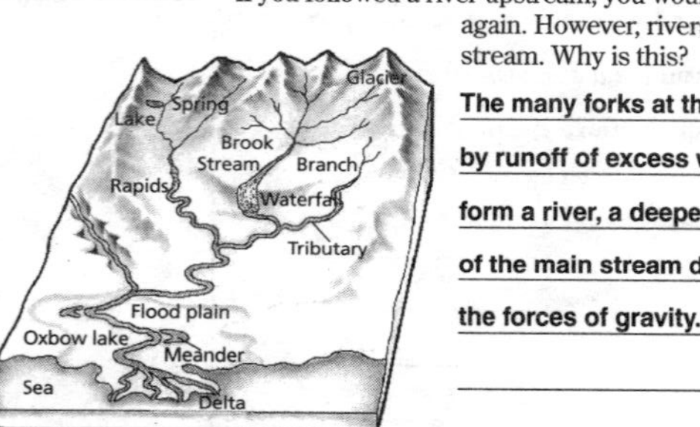


**The many forks at the headwaters of a river are caused by runoff of excess water. As the waters accumulate to form a river, a deeper and wider valley forms. The waters of the main stream do not fork downstream because of the forces of gravity.**

### 2. How Good a Geologist Are You?

To find out, take the following test. A list of observations is given below. Give one or more possible explanations for each one. Check your answers with your teacher to find out how you did.

| | |
|---|---|
| Super Geologist | 6 or 7 |
| Geologist | 4 or 5 |
| Future Geologist | 3 or less |

**Sample answers:**

**a.** Deep scratches are discovered in the surface of highly polished bedrock.

**During the last ice age, glaciers traveled here from another country, depositing rocks in your backyard.**

**b.** A perfectly preserved body is found in ice several kilometers from where it disappeared 75 years before.

**The body was frozen inside a glacier and moved along with the glacier as it flowed.**

**c.** Over a period of 5 years, a beach disappears.

**Several causes might explain the disappearance of a beach, but tidal encroachment and erosion from wave action are probably the major causes.**

Illustrations also on page 520 of your textbook

Chapter 26





Name ______________ Date ________ Class ________



**d.** A crack large enough to put your foot into runs the length of a huge boulder lying next to the highway.

**The boulder was probably deposited by a glacier. The crack might be the result of weathering from ice wedging into cracks or of blasting during highway construction.**

**e.** River water is salty many kilometers inland from the ocean.

**River water could be salty many kilometers inland due to river valleys being flooded when the ocean rises above previous levels. This is called the estuary effect.**

**f.** A huge granite boulder stands alone at the edge of a lake.

**As a glacier moved across the area, the lake could have been gouged out, and the boulder could have been deposited.**

**g.** A well goes dry in mid-August.

**The water table is the surface of the ground water. If the ground water is reduced, the water table drops below the level of the well, making it go dry.**

### 3. Threats to Ground Water

Because of your research skills, a journalist has hired you to do background research for a magazine article. Find out why the ground water in parts of the United States may be unsafe to drink. As you do your research, write down information that the journalist could use in the article.

**About 90 percent of the Earth's unfrozen fresh water is stored as ground water. Ground water may become contaminated by salt water, sewage, or pollutants. Pollutants include pesticides, fertilizers, and leaks from hazardous-waste dumps. Encourage students to find information about solutions to the problem of contaminated ground water and about private organizations and government agencies concerned with the problem.**

Name ______________ Date ________ Class ________

Chapter 26 Review Worksheet, continued

**4. Wasting Away?**

If streams and rivers are such effective agents of erosion, why have they not reduced the entire United States to sea level?

**While erosion carries away weathered material, new soil is constantly forming, and areas of the Earth's crust are uplifted. New crust is also formed by volcanic activity and by magma filling the gaps between diverging tectonic plates.**

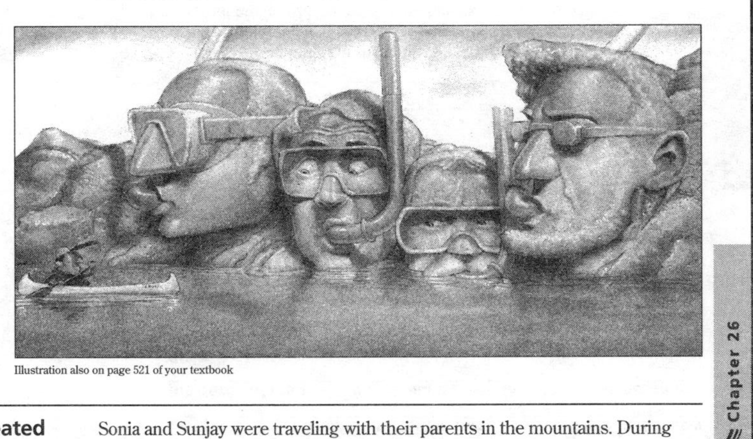

Illustration also on page 521 of your textbook

**5. A Heated Debate Over Ice**

Sonia and Sunjay were traveling with their parents in the mountains. During their trip they visited a glacier and spent some time walking around on it. An argument broke out when Sonia said, "This glacier is moving, you know." "No way," replied Sunjay, "I can't see or feel it move." Who won the argument? Why?

**Sonia was correct in saying that the glacier was moving. By definition, glaciers are moving masses of ice, although the flow of the ice is very slow.**



Chapter 26



Name ______________ Date ________ Class ________

Chapter 26 Assessment

## Word Usage

1. Using what you know about the movement of rivers across land, put the following terms in sequential order: *mouth, delta, river,* and *ocean.* Then write one or two sentences describing your sequence.

**Sample sequence: river, mouth, delta, ocean**

**The point where a *river* and the *ocean* meet is known as the *mouth* of the river. This is where the river dumps much of the sediments that it has been carrying, forming a body of land known as a *delta.***

## Correction/ Completion

2. The following sentences are incorrect or incomplete. Your challenge is to make them correct and complete.

**a.** Soil that allows large amounts of water to pass through it has a high porosity.

**Soil that allows large amounts of water to pass through it has high *permeability.***

**b.** The upper surface of the ground-water zone is known as the **water table**.

## Short Response

3. Put the sentences below in order by numbering them from 1 to 6.

**3** **a.** The gullies run together and form channels.

**5** **b.** The river carries more water and begins eroding its valley as well as its bed.

**1** **c.** A huge block of flat land is slowly lifted above sea level.

**4** **d.** Tributaries cut downward to form V-shaped valleys.

**2** **e.** Rainwater carves out gullies or ditches.

**6** **f.** The river meanders as it erodes sideways.

Name ______________ Date ________ Class ________



CHALLENGE 1

**Short Response**

4. Give a possible explanation for each of the following observations:

a. After a heavy rain, the river is brown.

**Sample answer: Runoff from a heavy rain carries dirt (or sediment) into the river, making the river appear brown.**

b. On a farm, there is a shallow well that runs dry if it has not rained for a few weeks, but there is a deep well that always has water in it.

**Sample answer: When there is little rain, the water table falls below the depth of the shallow well, but it never falls below the depth of the deep well.**

c. Many of the rivers in Maine flow more swiftly than many of the rivers in Iowa.

**Sample answer: Maine is more mountainous than Iowa and has more glaciers and snow, which melt and create swiftly flowing rivers.**

CHALLENGE 2

**Graph for Interpretation**

5. A large area surrounding a stream has been heavily developed with homes, parking lots, and roads. Examine the graph below, which shows the amount of runoff into the stream before the area was developed and after the area was developed. Then answer the questions on the next page.

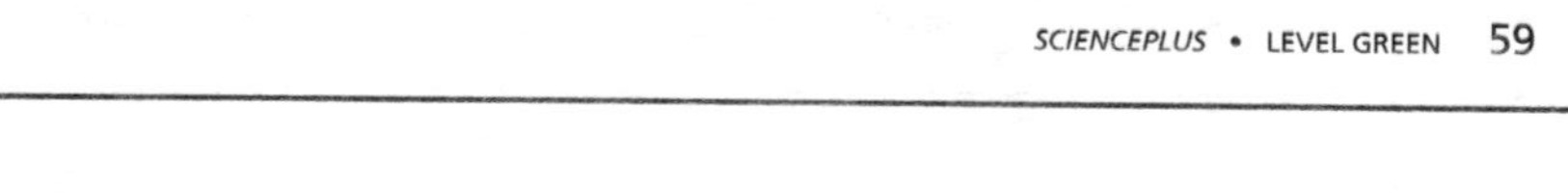


Chapter 26





Name ______________ Date ________ Class ________



a. Which curve shows the amount of runoff after the area was developed? Explain.

**Sample answer: Curve A shows the amount of runoff after development. Before the area was developed, there were woods and fields that would absorb a lot of rainfall. Only a fraction of the rainfall became runoff, and the runoff took a while to reach the stream. Now that the area is heavily developed, not as much water is absorbed. Instead, it runs quickly into the stream.**

b. Are floods more likely to occur before or after development? Explain.

**Sample answer: Floods are more likely to occur after development because the amount of runoff is increased.**

Name ______________ Date ________ Class ________

**Unit 8**
Unit Activity Worksheet

## Geology Match-Up

Match these geologic words and phrases as you conclude Unit 8.

Place a ruler on an imaginary line connecting each phrase on the left with the best match on the right. Each line will cross a letter and a number. The number tells you where to put the letter in the line of boxes at the bottom of the page. What is the message in the boxes?

|  |  |
|---|---|
| a natural change 9 | outwash |
| any large movement of the Earth's crust | moraines |
| small rock fragments dropped by a glacier | downslope movements |
| glacial evidence | weathering |
| a crack in the sidewalk | tectonic movements |
| coastlines are shaped by these | weathered tombstone |
| avalanches and mudslides are called this | old rivers |
| excessive runoff may cause this | tides |
| low porosity 11 | clay |
| rivers that meander | theory of continental drift |
| breakwaters prevent this | delta |
| flooded coastal valley | river channels |
| Alfred Wegener | glacial movement |
| fan-shaped wedge of sediment | estuary |
| gravity acting on ice | sand |
| has high permeability 12 | flooding |
| gullies run together to form these | erosion |

| 1 | 2 | 3 | 4 | | | 5 | 6 | 7 | 8 | | | 9 | 10 | | | 11 | 12 | 13 | 14 | 15 | 16 | 17 |
|---|---|---|---|---|---|---|---|---|---|---|---|---|---|---|---|---|---|---|---|---|---|---|
| Y | O | U | R | | | W | O | R | K | | | I | S | | | P | E | R | F | E | C | T |





Name ______________ Date ________ Class ________

**Unit 8**
Unit Review Worksheet

## Making Connections, page 522

**The Big Ideas**

In your ScienceLog, write a summary of this unit, using the following questions as a guide:

1. What is geology all about? (Ch. 24)
2. What does the expression "the present is the key to the past" mean? (Ch. 24)
3. How do geologists use inferences and observations? (Ch. 24)
4. What is some evidence that supports the theory of continental drift? (Ch. 24)
5. What are the main features of the Earth's interior? (Ch. 24)
6. Use your knowledge of the structure of the Earth to explain why the floor of the Atlantic Ocean is growing wider. (Ch. 24)
7. What are some of the agents of weathering? (Ch. 25)
8. What role does water play in the weathering process? (Ch. 25)
9. What role does gravity play? (Ch. 25)
10. What is some evidence for widespread glaciation in the past? (Ch. 26)

**A sample unit summary is provided on page 522 of the Annotated Teacher's Edition.**

**Checking Your Understanding**

1. A famous saying about geology is written in code below. Here is how to crack the code. One letter simply stands for another.

CHKOOKLPANEO

GLOSSOPTERIS

Use the example to get started. Can you decipher this famous saying?

PDA LNAOAJP EO PDA GAU

**THE PRESENT IS THE KEY**

PK PDA LWOP—FWIAO DQPPKJ

**TO THE PAST—JAMES HUTTON**

Name ______________ Date ________ Class ________



2. Imagine that all tectonic activity suddenly came to a grinding halt. What would the Earth be like in 10 million years? 100 million years? 1 billion years? Continue your answer in your ScienceLog if necessary.

**Answers will vary. Sample answer: If all tectonic activity were to suddenly stop, the continents would probably stop moving. As a result, their locations would probably remain much as they are now, even 1 billion years from now.**

3. Most words have more than one meaning. Read each sentence below and explain the meaning of the underlined term. Then write a sentence to show that you understand the geological or scientific meaning of the underlined term as it was used in this unit.

**Example:** Betty and Sergei were in the runoff race.
Heavy rains can cause great damage due to runoff.

a. He is in big trouble for having his hand in the till.
**Hand in the *till* means stealing from a cash drawer. Geological meaning: The till deposited by the glacier consists of many different types of rock.**

b. How are you weathering the storm?
***Weathering* the storm suggests surviving a difficult situation. Geological meaning: The weathered landscape had only large boulders in it that could not be carried off by the wind.**

c. Why not meander over to my house after supper?
***Meander* here means a slow, wandering walk. Geological meaning: Many meanders had developed in the old river.**





Name ______________ Date ________ Class ________



d. Her election win was a landslide victory.
***Landslide* here means a victory by a large majority. Geological meaning: The earthquake triggered the landslide.**

e. She received an avalanche of phone calls after she won the trophy.
***Avalanche* here means a great number. Geological meaning: The avalanche was started on purpose by an explosion.**

f. The baseball missed home plate by a country mile.
**Home *plate* is home base in a baseball game. Geological meaning: The Earth's crust is broken up into many plates.**

g. Is a water table like a water bed?
**A *table* is a piece of furniture that consists of a flat, horizontal top set on legs. Geological meaning: The water table is the flat, horizontal level below which the ground is saturated with water.**

Illustrations also on page 523 of your textbook

Find other examples of words used in this unit that have both a scientific and a common meaning, and record them in your ScienceLog.

4. How do you think the process of erosion would be affected if
a. the force of gravity were doubled?
**Doubling the force of gravity would greatly increase erosion.**

Name ____________ Date ________ Class ________

**Unit 8 Review Worksheet, continued**

**b.** the average global temperature went up by 5°C?

**This might cause more moisture to accumulate in the atmosphere, which would cause greater weathering and erosion.**

**c.** the atmosphere vanished?

**Weathering and erosion would greatly decrease because there would be no wind or water.**

**d.** the boiling point of water suddenly changed to 10°C?

**Except in areas with temperatures below 10°C, there would be no natural bodies of water because they would have boiled away. This would decrease the amount of erosion.**

**e.** there were no gravity?

**Erosion would cease because gravity is the force behind all erosion.**

**f.** water froze at 25°C?

**Most of the Earth's surface would be covered by ice. This would increase the amount of erosion.**





Name ____________ Date ________ Class ________

**Unit 8 Review Worksheet, continued**

**g.** water were four times heavier (for a given volume)?

**Erosion would greatly increase due to the increase in the force of the water. (That is, the weight of the water would result in increased pressure.)**

**5.** concept map Make a concept map using the following terms or phrases: *waves, ice, gravity, erosion, wind,* and *flowing water.*

**Sample concept map:**



Illustration also on page 523 of your textbook

Name ______________ Date ________ Class ________

## Unit 8
End-of-Unit Assessment

### Word Usage

1. Compare the *porosity* and *permeability* of sand and garden soil. Write a sentence to show that you understand these terms.

   **Sample answer: The *porosity* (percentage of empty space) and the *permeability* (the ability to transmit a fluid, such as water) of sand are generally greater than that of garden soil.**

2. For each pair of words below, show the connection between the words by using them in a sentence.

   a. *water table* and *lakes*

   **Sample answers: *Lakes* form where depressions in the Earth's surface penetrate the *water table*.**

   b. *deposition* and *deltas*

   **When sediment-carrying rivers meet the ocean, *deposition* occurs, and *deltas* are formed.**

   c. *continental drift* and *fossils*

   ***Continental drift* explains why *fossils* from tropical lands have been found in Antarctica.**

### Correction/ Completion

3. The following statement is either incorrect or incomplete. Your job is to make it correct and complete.

   During an ice age, **glaciers** advance over the surface of the Earth. When they stop, a ridge of deposit called a **terminal moraine** is formed.

### Short Responses

4. Read the observations and inferences below. Decide if each inference is reasonable. If not, write a reasonable inference on the next page.

| Observation | Harold's inference |
|---|---|
| a. A ball lies by a broken window. | The ball caused the window to break. |
| b. The Mississippi River floods in the spring. | Melting snow and spring rains cause the floods. |
| c. A river has carved a narrow channel through a rugged mountain valley. | The river dates back to the beginning of time. |







Name ______________ Date ________ Class ________



| Observation | Harold's inference |
|---|---|
| d. Granite boulders and rocks are found in farmland 200 km from mountains made of granite. | These boulders and rocks have always been there. |

**Inferences (a) and (b) are reasonable. (c) Over thousands or millions of years, the river eroded a valley through the mountains. (d) The boulders and rocks were carried to their current location by glaciers.**

5. Geologists use their observations to make inferences about what must have taken place in the past. Use your "geo-logic" to draw inferences about each of the following situations:

   a. Some of the rock layers in a cliff contain fossils of fish, oysters, and clams.

   **Sample answer: The area was once under water.**

   b. You find an exposed layer of rock whose surface bears deep grooves and appears to have been polished.

   **Sample answer: The area was once covered by glaciers.**

### Illustrations for Interpretation

6. Which of the figures below would show the fastest rate of weathering, and why?

   a. b. c.

   **Figure (c) would erode the fastest because it has a greater surface area exposed to the elements of erosion.**

Name ______ Date ______ Class ______



7. Describe what is happening to the river in the diagram below.

a. b. c.

Sample answer: As this river ages, erosion occurs sideways, not downward. At the same time, deposition occurs on the opposite bank. This causes the river to meander. Eventually, the river may cut off a meander by eroding through it, leaving an oxbow lake.

**Illustration for Correction or Completion**

8. Sketch what you think the cliff shown below might look like after 100 years if the cliff were made of (a) hard granite, (b) soft sandstone, and (c) soil and clay. Include a written description with each sketch if you wish.

Cliff face
Crack
Cave
Sea level

**Sketches should show the following: (a) The granite cliff should look much the same, perhaps with the cave enlarged slightly and a few rocks broken off. (b) The soft sandstone cliff should have worn away significantly with a substantially enlarged cave. (c) The cliff of soil and clay would have eroded far back, possibly beyond the edge of the illustration. It may have begun to form a slope.**





Name ______ Date ______ Class ______



CHALLENGE

**Short Response**

9. How is a glacier like a bulldozer?

Sample answer: As a glacier moves forward, it pushes huge amounts of soil and rock along with it. When the glacier stops, it dumps the soil, rock fragments, and boulders, much like a bulldozer does.

**Short Essay**

10. Factors that affect how fast weathering and erosion occur differ from place to place. Name as many differences as you can between the speed and amount of weathering in the following pairs of locations:

**a.** a mountain slope in the Sierra Nevadas and a beach in Florida

Both locations would weather rapidly. In the Sierra Nevadas, a continuous cycle of freezing and thawing, as well as daily exposure of the rock to a wide range of temperatures, would cause cracking and other types of weathering. On the beach in Florida, the constant pounding of waves would cause the shoreline to weather rapidly. Chemical forms of weathering, however, would be far more significant on the beach than in the Sierra Nevadas. This is because the warmth and high humidity of Florida contribute to chemical weathering more than the cold, dry climate of the Sierra Nevadas does.

Name ______________________ Date __________ Class __________



**b.** the moon and a ranch in Brazil

**Sample answer: The ranch in Brazil is susceptible to the weathering forces of wind and water. If trees are cut down on the ranch and livestock are allowed to overgraze, the rate of erosion would be high. In addition, chemical weathering would be fairly significant because of the warm, moist climate in Brazil. The moon, on the other hand, has no atmosphere, and would thus not be prone to these types of weathering. However, because there is no atmosphere, the moon is not protected from the effects of falling meteors and other debris from space. These objects contribute greatly to the erosion of the moon's surface.**

CHALLENGE 2

## Illustration for Interpretation

**11.** The sketch below represents a new house built next to a hill. Mark the positions on the diagram where the owner may have some cause for concern. Then explain why the owner should be concerned.

**Sample answer: The owner should be concerned with the hillside to the left of the house because of the possibility of either slumping or sliding. Runoff from the hill could also be a problem. The nearness of the basement to the water table could be a cause for concern. A slight rise in the water table could flood the basement.**





Name ______________________ Date __________ Class __________



## Short Essay

**12.** Arthur's family decided to build a swimming pool in their backyard. Bulldozers came in late fall and excavated the hole, but soon the temperature fell below freezing, and construction stopped. This gave Arthur an idea for a project on weathering and erosion. Write a brief account of the investigation Arthur might have done during the fall, winter, and spring.

**Sample answer: The purpose of the investigation could be to measure the effects of weathering and erosion on freshly exposed soil and rock. The study could start with observations of the freshly exposed rock and soil. These observations could be repeated at regular intervals or after a major event, such as a heavy rainfall or a hard freeze, and the changes noted. Arthur could conclude that weathering and erosion begin immediately after a fresh surface is exposed, that freezing and thawing have a major effect on exposed rock, and that rainfall, especially heavy rainfall, is a major cause of erosion.**

HRW material copyrighted under notice appearing earlier in this work.



HRW material copyrighted under notice appearing earlier in this work.

Name ______________ Date ________ Class ________

Activity Assessment, continued

## Observation Chart

| Sample | Description of sample before erosion and prediction | Description of sample, water, and jar after erosion | Erosion resistance |
|---|---|---|---|
| 1. Chalk | **Sample answers: The chalk is easy to scratch. Its fine grains will probably wear away quickly and completely.** | **The chalk has completely disappeared, giving a milky appearance to the water. Some of the small rock particles have adhered to the jar.** | **Lowest** |
| 2. Granite | **The granite is hard to the touch and does not wear away when scratched. It will probably not erode much.** | **The granite has eroded very little. The water is still clear, with only a few pieces of rock at the bottom of the jar.** | **Highest** |
| 3. Sandstone | **The sandstone is composed of several larger grains joined together with numerous smaller grains. The smaller grains will probably erode away, leaving behind the larger grains.** | **The water has many small particles floating in it. Some of these particles are sticking to the bottom and sides of the jar. There is a large chunk of rock that has not eroded.** | **Middle** |





Name ______________ Date ________ Class ________

Unit 8 SourceBook Activity Worksheet

## A Rock-Cycle Simulation

Complete this activity after reading pages S180–S181 of the SourceBook.

| Your goal | to learn about the different stages of the rock cycle by simulating the formation of sedimentary, igneous, and metamorphic rocks | Safety Alert! |
|---|---|---|

### You Will Need (per group)

- newspaper
- 2 to 3 each of red, green, blue, and yellow crayons
- a small pencil sharpener or scissors
- heavy-duty aluminum foil
- 2 large, heavy books, such as a dictionary or encyclopedia
- a watch with a second hand
- a metric ruler
- an aluminum pie plate
- oven mitts
- a hot plate

### What to Do

1. Form into groups of three or four students.
2. Cover your desktop or work area with two or three sheets of newspaper. Shave each crayon into fragments using a small pencil sharpener or a pair of scissors, keeping the crayon shavings separated into four piles according to color.
   **Caution: If using the scissors, be very careful not to cut yourself!**
3. What did the whole crayons represent in this rock-cycle simulation? What might the crayon shavings represent? the pencil sharpener or scissors?
   **The whole crayons represent rocks, the shavings represent rock fragments, and the pencil sharpener or scissors represent weathering agents such as the sun, rain, wind, and frost.**
4. Tear off a sheet of aluminum foil about 20 cm × 20 cm. You may need to trim the sheet with scissors to get the proper dimensions. Fold this sheet in half. Place the foil on the newspaper, opening it so that you can see the crease in the middle.
5. Sprinkle the crayon shavings of one color in the center of the flat side of the foil so that they cover an area of approximately 4 cm × 4 cm to a depth of 0.5 cm. Then apply the other colors of crayon shavings, creating layers of color.
6. What might these layers of crayon shavings represent?
   **Accumulated layers of rock fragments or sediment deposition**

Name ______________________ Date __________ Class __________

SourceBook Activity Worksheet, continued

7. Fold the top half of the foil over the crayon layers. Place this foil package between two heavy books, and apply light pressure for 2 seconds. Remove the foil package from the books and open it.

8. What has occurred? What kind of rock did you simulate? What do the books represent in this simulation?

**The crayon shavings have been pressed together. The formation of sedimentary rock has been simulated. The books represent the weight of the accumulated layers of sediment.**

9. Place the crayon "rock" back in the foil, and put the foil between the two books again. This time both you and a partner press as hard as you can against the books for 1 minute. Remove the package from between the books and open it.

10. What happened to the crayon "rock layers" this time? What kind of rock has been simulated? How are the books used differently in this simulation?

**The layers are more compressed and have fused together. Metamorphic rock has been simulated. The books act differently because they apply more pressure to the crayon shavings. This represents intense heat and pressure.**

11. Put the aluminum pie plate on the hot plate. Wrap the crayons in the foil again and place the package in the pie plate. Turn the hot plate on.

12. What do you think is happening to the crayon "rock"? **Caution: Do not touch the hot foil package!** What does the hot plate represent?

**The crayon shavings are melting. The hot plate represents the intense heat of the Earth's interior core.**

13. Turn off the hot plate and allow the crayon "rock" to cool and harden completely. What type of rock did you simulate?

**Igneous rock is simulated by the melting and cooling of the crayon shavings.**

14. How is this simulation process different from the rock cycle that actually occurs in nature?

**Sample answer: It takes millions of years for rocks to go through just one part of the rock cycle, while all stages were represented in a very brief period of time in this simulation.**





Name ______________________ Date __________ Class __________

Unit 8 SourceBook Review Worksheet

SourceBook

## Unit CheckUp, page S205

### Concept Mapping

The concept map below illustrates major ideas in this unit. Complete the map by supplying the missing terms. Then extend your map by answering the additional question below.

**Sample concept map:**

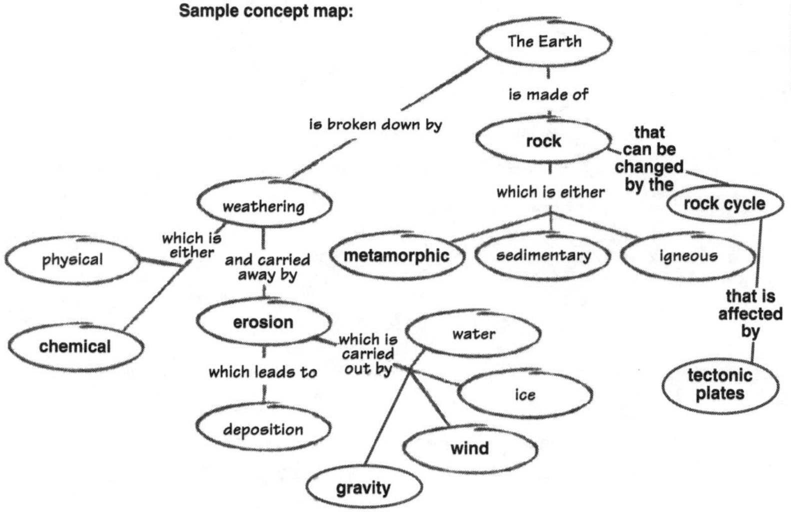

Where and how would you include the terms *rock cycle, gravity,* and *tectonic plates?*

Name ____________ Date ________ Class ________

SourceBook Review Worksheet, continued

## Checking Your Understanding

Select the choice that most completely and correctly answers each of the following questions.

1. Which type of rock is formed from molten material?
   a. (igneous) b. sedimentary
   c. metamorphic d. magma
2. Which is true of the rock cycle?
   a. (Any one rock type can be changed into any other.)
   b. Sedimentary rock always changes into metamorphic rock.
   c. Igneous rock always turns into sedimentary rock.
   d. The change from one rock type to another takes several decades.
3. Which would NOT be an example of physical weathering?
   a. A sidewalk cracking in the summer sun
   b. A hole being scoured in a boulder during a sandstorm
   c. (A drop of rainwater dissolving a tiny bit of the cement holding sandstone together)
   d. A boulder being split apart by the roots of an oak tree
4. The faster a stream flows,
   a. the less likely it is to cause major erosion.
   b. the larger the delta it will deposit.
   c. the more likely it is to have a flood plain.
   d. (the greater its erosive ability.)
5. The abyssal plains
   a. are often swept by strong currents.
   b. are never found far from land.
   c. (are sites of little activity.)
   d. are home to many plants and animals.

## Interpreting Photos

The photo on page S206 of your textbook shows an unusual land feature that has been formed by the action of erosion on its surface. Using what you have learned in this unit, which type of erosion—gravity, streams, glaciers, wind, or waves—might have been responsible for producing this landmark? Explain.

**This is a photograph of the Matterhorn, a famous peak that rises 4478 m above sea level in the Pennine Alps between Italy and Switzerland. The smooth, steep sides of the peak were formed by the action of glaciers that scoured its sides, creating a bowl-shaped hollow called a *cirque* just below the peak.**







Name ____________ Date ________ Class ________

SourceBook Review Worksheet, continued

## Critical Thinking

Carefully consider the following questions, and write a response that indicates your understanding of science.

1. In what major ways do wind erosion and water erosion differ?

   **Water erosion moves both large and small particles. Wind erosion moves only very fine-grained particles.**

2. Eventually, the Earth may lose so much heat that its interior becomes solid. What effect might this have on the Earth's landforms?

   **If the Earth becomes solid, then tectonic plates will stop moving and landforms will stop being created. Eventually the planet's landforms will wear down to a more or less level plain near sea level.**

3. Weathering does not occur evenly throughout the world. Which types of climates probably experience the fastest weathering? the slowest? Why?

   **Very warm and moist climates, such as those found in rain forests, probably experience the fastest rates of weathering. Very cold and dry climates, such as those found near the poles, probably experience the slowest rates of weathering. Warm, wet climates experience faster weathering because the chemical processes that break things down work faster in warm conditions than in cold and because water is a major agent of chemical weathering. The reverse is true of cold and dry climates.**

SourceBook

Name ______________ Date ________ Class ________

SourceBook Review Worksheet, continued

**4.** Tests have shown that, as you move away from the center of the mid-ocean ridges, the rocks become steadily older. How does this evidence support the idea that the Earth's crust moves?

**This evidence suggests that rock is formed at the mid-ocean ridges and then carried away, presumably by the motion of the Earth's surface.**

**5.** Lakes are usually short-lived features, geologically speaking. Explain why this is so, using the concepts of erosion and deposition.

**Lakes are low places on the surface. Furthermore, they are usually fed by streams or rivers. Because streams and rivers generally contain a certain amount of rock and soil and because sediments tend to settle in low places, the lakes are eventually filled with the products of weathering.**

## Portfolio Idea

As you know, the Earth is estimated to be about 4.5 billion years old. This is a span of time most people find very difficult to grasp. Design a scenario appropriate for helping an eight-year-old child to understand this span of time, using the major geologic events shown on page S184 of your textbook. For example, you might say "If the Earth's history were compressed into one year, oceans began forming on February . . ." and so on. Don't use this example. Be imaginative, but be accurate. Record your scenario in your ScienceLog.

**Student examples will vary. Answers should reflect an appropriate analogy of the geologic time scale.**





Name ______________ Date ________ Class ________

**Unit 8** SourceBook Assessment

SourceBook

1. The Earth's surface is constantly changing.
   a. (true) b. false
2. Rocks both on the Earth's surface and under it can undergo weathering.
   a. true b. (false)
3. Soil contains ______, which is the decaying remains of dead plants and animals.
   a. clay b. (humus) c. gravel d. dust
4. Soil that is high in humus is fertile.
   a. (true) b. false
5. Which of the following does not cause a change in the composition of rock that is being eroded?
   a. (physical weathering) b. chemical weathering
6. If you find rocks in the bottom of a river bed that are smooth with almost no sharp edges, you are probably observing an example of what type of weathering?
   a. exfoliation b. expansion c. (abrasion) d. frost action
7. Piles of broken rock pieces at the bottom of a cliff or steep slope are called
   a. slump. b. creep. c. (talus.) d. channels.
8. The most common form of erosion due to gravity is
   a. slump. b. (creep.) c. talus. d. channels.
9. The Grand Canyon is an example of
   a. erosion by gravity. b. (erosion by streams.) c. erosion by glaciers. d. erosion by wind.
10. The Great Lakes were formed by
    a. valley glaciers. b. (continental glaciers.)
11. A flood plain contains very poor soil and is not suitable for farming.
    a. true b. (false)